Springer Series in Measurement Science and Technology

The Springer Series in Measurement Science and Technology comprehensively covers the science and technology of measurement, addressing all aspects of the subject from the fundamental principles through to the state-of-the-art in applied and industrial metrology, as well as in the social sciences. Volumes published in the series cover theoretical developments, experimental techniques and measurement best practice, devices and technology, data analysis, uncertainty, and standards, with application to physics, chemistry, materials science, engineering and the life and social sciences.

More information about this series at http://www.springer.com/series/13337

Leslie Pendrill

Quality Assured Measurement

Unification across Social and Physical Sciences

Springer

Leslie Pendrill
Partille, Sweden

ISSN 2198-7807 ISSN 2198-7815 (electronic)
Springer Series in Measurement Science and Technology
ISBN 978-3-030-28694-1 ISBN 978-3-030-28695-8 (eBook)
https://doi.org/10.1007/978-3-030-28695-8

This Springer imprint is published by the registered company Springer Nature Switzerland AG
The registered company address is: Gewerbestrasse 11, 6330 Cham, Switzerland

Preface

Ever-increasing demands for comparability and reliable risk assessment for sustainable development in the widest sense place corresponding demands on the measurements on which important decisions are based, in both traditional and new areas of technology and societal concern. International consensus about metrological traceability and uncertainty—both conceptually and in implementation—which ensure comparability and risk assessment has however yet to be achieved in every field.

This book shares several of these measurement goals and challenges with other titles in the Springer Series in Measurement Science and Technology but complements the approaches of others by presenting measurement in the context of conformity assessment. This allows measurements to be anchored in relevance and interest for third parties. It turns out also to provide the key to a unified presentation about quality-assured measurement across the social and physical sciences.

The whole layout of this book about quality-assured measurement across the social and physical sciences reflects our approach, where each chapter treats successive steps in a quality assurance loop for conformity assessment aimed at keeping product on target. The reader is encouraged at the end of most chapters to fill in a template for their product of choice at successive stages in the quality assurance loop, so that by the end of the book a complete evaluation of quality assurance for that product should be at hand. A worked-out example dealing with weighing and the measurement of perceived prestige when drinking coffee is provided as a running case study through the book.

Physical quantities possess some remarkable properties which enable correspondingly remarkable possibilities for metrologically traceable measurements; not only can the results of measurements of a particular quantity be compared, but also measurements of different quantities can show a degree of comparability. The remarkable properties in physics are of course not necessarily shared by quantities in other disciplines, and avoiding what has recently been termed ‘physics envy’ is an issue of course in the context of the present book, aiming to give a unified presentation of quality-assured measurement across the social and physical sciences.

The historic division over measurement between physicists and psychologists in the twentieth century and the naissance of a unification from the turn of the millennium are mentioned in Rossi's book in this series. In this book, we consider how far the engineering approach of measurement system analysis can be extended to other fields. The key to our approach to describing measurements in the social sciences is to treat the human responder herself, rather than the questionnaire, as the 'instrument' at the heart of the measurement system. For the physical scientist and the metrologist, our approach brings novelty, by suggesting reversing the traditional handling of measurement uncertainty—starting not with standard deviations but with 'quandary' when making decisions. The commonality between physical and social measurement (and qualitative estimations more generally) is first reached when one recognises that the performance metrics of a measurement system (how well decisions are made) are the same concept in both. The Rasch psychometric approach, as an item response theory from the 1960s but with roots in the 'counted fraction' studies of early twentieth century pioneers such as Tukey and Pearson, is shown to play a special role in the measurement process as a concatenation of observation and restitution.

Some social scientists remain sceptical about finding sufficient objectivity in human-based measurements to allow the establishment of comparability through metrological traceability. A Rasch approach to the enduring beauty of fine art or perceived happiness (or other pleasing patterns and degrees of order or symmetry) would, we believe, provide separate measures of the (albeit noisy) individual preferences of different persons and the intrinsic ability of, respectively, Leonardo's paintings to stimulate pleasure or a particular activity of daily living to invoke health and happiness.

The reader may find some original material in this book (always a risky claim, especially today when so much is researched and available on the Internet), including new insight into measurement units in quantum mechanics (coinciding with a major revision in 2019 of the SI in terms of fundamental constants), as well as the role of entropy in explaining both key aspects of metrology—comparability and uncertainty—in terms of symmetry and the amount of information. While not enjoying access to universal units of measurement as in physics, we attempt to demonstrate how in the social sciences one can indeed establish recipes for traceability analogous to certified reference materials in chemistry, by defining measurement units with construct specification equations.

Ultimately, the aim of any measurement should be to achieve *effectiveness*, meaning not mere passive observation but 'changing conduct' in terms of relations between signs of communication having to do with actively 'improving' the entities they stand for.

Partille, Sweden Leslie Pendrill

Acknowledgements

Much of the contents of this book results from educational material developed since the 1980s in courses ranging from industrial measurement training, online measurement handbooks (metrology.wordpress.com), to university postgraduate education where I need to thank generations of students who have acted as 'guinea pigs' in the 'honing' process essential to any education.

I have had the benefit of working continuously over the decades in the service of the national metrology institutes (NMI)—for most of my career as Head of Research at the Swedish Metrology Institute—and have chaired a number of European and international metrology organisations. Apart from national funding, the support of a long line of projects sponsored by the European Commission has been essential—particularly the Measuring the Impossible initiative of the 2000s and most recently the NeuroMET projects[1] of the European Metrology Programme for Innovation and Research (EMPIR).

In addition to NMI colleagues, such projects have forged liaisons with prominent collaborators, including social scientists striving for a unified metrological understanding and whose names may be found among authors of this Springer Series in Measurement Science and Technology as well as in the Acknowledgements and co-authors of many of my publications.

[1]The NeuroMet project has received funding from the EMPIR programme co-financed by the Participating States and from the European Union's Horizon 2020 research and innovation programme.

Contents

Chapter 1
Measurement Challenge: Specification and Design

'*Measurement is not an end in itself* ...' might seem to be a paradoxical way of introducing a book about measurement. But measurement is important to the majority since it gives objective evidence on which to base decisions. The need for quality-assured measurement has evolved and widened over the centuries:

Measurement for trade and ownership, with ancient roots in buying, selling, bartering, exchange, map-making on so on, is still relevant in today's globalised world. Measurement in science: from the seventeenth century, experiments became the arbiter of fledgling theories about the nature of the physical world. Measurement in social science: from the nineteenth century, psychophysics was followed later by psychometrics and econometrics. With the industrial revolution in the mid-nineteenth century and throughout the twentieth century, quality assurance emerged in manufacturing/production, where product quality depends on measurement quality. Better products in turn lead to enhanced competitiveness, less waste, sustainable production, greater efficiency...To these well-established applications of quality-assured measurement have been added typical twenty-first century concerns: Information and communication technologies (ICT) depend on reliable synchronisation and spectrum allocation. The Environment: objective measurement provides support to monitoring, sustainability, conformity assessment and risk and cost analyses in the face of this grand challenge. Finally, in health, safety and security, reliable measurements are of course essential in diagnosis and medical treatment, robust ICT of health and organisation data, reliable nuclear fuel data and so on. What is evident from this albeit incomplete and cursory list is that quality-assured measurement remains a topic of burgeoning and increasingly multidisciplinary interest.

This book will deliberately not follow the many exposés already in the literature which set the subject of measurement in an historical context, however evocative that may be. In-depth accounts of how the ancient Egyptians or bold navigators struggled when pioneering quality-assured measurement will be mostly left to others to present.

L. Pendrill, *Quality Assured Measurement*, Springer Series in Measurement Science and Technology, https://doi.org/10.1007/978-3-030-28695-8_1

There is a need and a challenge to formulate a unified view of measurement. To this end, most of this first chapter of the book—as well as the last—will paradoxically *not* deal directly with measurement, but rather the objects—products, services, concepts... and their characteristics—which are the concern of many people, who then ask metrologists to measure them. The first and last chapters thus provide object-related 'bookends'—supporting a description of quality-assured measurement which is the central issue.

Presenting measurement in relation to objects will allow measurements to be anchored in relevance and interest for third parties. As it turns out, the approach also provides the key to a unified presentation about quality-assured measurement across Social and Physical Sciences where objects are probed by Man as a measurement instrument.

1.1 Processes of Production and Measurement

An illustrative story of the importance of measurement when regarding 'quality' in terms of meeting expectations is an account (Swyt 1992) of how a slight difference in dimension tolerancing of car door frames made a Japanese car model, with a 1 mm tolerance, so much more sellable than the otherwise comparable American car, where the door frame tolerance was 2 mm instead. Engine power, acceleration, plush interior fittings, etc. meant relatively little in determining the competitive advantage of one car model over the other, compared with the experiences of potential customers when the car door was to be opened and closed. The customers were of course unaware of the specific manufacturing tolerances but did at first-hand experience clearly the impact of these in the way product is perceived. Note how the different kinds of properties—physical, sensory, psychological, social and economic—blend in this simple narrative of the interplay between product and measurement.

A rather different example of the importance of measurement is its role in providing, through experiment, tests of the validity of new ideas in physics. Feynman (2013) makes a strong case for measurement: 'The principle of science, the definition, almost, is the following: *The test of all knowledge is experiment.* Experiment is the *sole judge* of scientific "truth."' Again, 'quality' has to do with meeting expectations.

The emergence of fields such as 'soft reliability' is a response to the fact that in recent years the majority of faults in consumer products (even for 'hard' products such as electronics and IT devices) are no longer purely technical, but are rather dominated by complaints such as 'it doesn't work' or 'I couldn't find the start button' (den Ouden et al. 2006). Such a shift has of course serious consequences, both for repair workshops at points of sale still geared to servicing technical faults, as well as production processes where traditional divisions between the workshop floor, design and customer relations and marketing have to be radically re-thought.

In all of these three examples about meeting expectations through better measurement, and in many other actual measurement challenges across the physical and social sciences, a major task is to ensure that processes—of both production and measurement—keep product on target. For most of this book about measurement, the 'product' will not be a car or a washing machine, but a measurement result. Although not as concrete as a product, a measurement result nevertheless can be considered the 'product' of a process in which results are 'manufactured' (Nilsson 1995). Fruitful analogies will be drawn between entity and measurement conformity assessment procedures and concepts throughout this book.

Since neither production nor measurement processes are perfect, assuring the quality of the product or other entity (process, phenomenon, etc.) will involve efforts to keep within tolerable limits the unavoidable, real or apparent dispersion in the product (entity[1]) value, either for measurements of a series of items or repeated measurements of one item.

Conformity assessment is a formal process with the specific aim of keeping product on target. In assessing conformity to specifications of a product, service or other entity, the intention of is (ISO 2018):

(a) to provide confidence for the consumer that requirements on products and services are met,
(b) to provide the producer and supplier with useful tools to ensure product quality,
(c) often essential for reasons of public interest, public health, safety and order, protection of the environment and the consumer and of fair trading.

In physics, when judging the truth of a scientific concept, one might test experimentally whether a certain phenomenon has a value above or below a predicted limit; or more generally if the latest results confirm or refute earlier observations.

Conformity assessment of measurement is often a prerequisite and will be so to say 'nested inside' the process of conformity assessment of product. Conformity assessment has a focus on determining actual entity errors: apparent dispersion due to limited measurement capability should normally be small. In the same way as quality assurance of products is recommended in all processes of manufacture—from defining to delivering product—an analogous quality assurance loop (Sect. 1.3 and Fig. 2.1) should be identified to keep measurements on target.

1.2 Measurement, Assessment, Opinions: From Quantitative Observations to Categorization

Whether evaluations (measurements, estimations and opinions) are indeed sound and/or objective at all in for instance the social sciences is still under debate (Mari et al. 2017; Sawyer et al. 2016; McGrane 2015). Quoting McGrane (2015) for instance: 'Psychological measurement needs to be de-abstracted, rid of operational

[1]Wherever possible, use will be made of the international vocabulary for conformity assessment in choice of terminology—see Sect. 1.4.

rules for numerical assignment and set upon a foundation of quantitative theory, definition and law. In the absence of such theoretical foundations, claims of measurement in the psychological sciences remain a methodological chimera.'

1.2.1 Major Challenges and Interdisciplinary Studies

There is a growing awareness that many of society's so-called 'big challenges'—global climate change, the aging population, etc. (this Chapter)—should be tackled with increased mobilisation across disciplinary boundaries between the social and physical sciences to reach an increased stringency, for example, in decision-making and conformity assessment. At the same time, social scientists express frustration over what they perceive as entrenched attitudes towards achieving these goals (Viseu 2015). Among other things, the lack of a common language or complementary methodologies creates considerable obstacles for cooperation between sociologists, physicists and others.

In pursuit of increased stringency in social measurements and decisions, Nelson (2015) invites us to:

- be realistic, recognise limitations when social measurements are to be quantified,
- avoid 'physical envy'.

As Nelson (2015) writes: 'The emphasis on quantification and mathematical theorizing that has been so successful in physics may not be so appropriate in sciences dealing with other kinds of subject matter.... Whereas physics can limit the subject matter it addresses so that such heterogeneity is irrelevant to its aims, for other sciences, this diversity or variability is the essence of what they study.' 'Expecting science to achieve physics-like, quantified precision that can allow us to optimize policies in domains as diverse as cancer treatment, industrial innovation, K-12 education, and environmental protection is a fantasy. Here we will need to focus on improving our processes of democratic decision making' (Nelson 2015).

Even from the neighbouring discipline of analytical chemistry, the metrological vocabulary developed mainly in physical terms is regarded as in need of revision. Quoting from the recently proposed VIN (International vocabulary for nominal properties, (Nordin et al. 2018)): 'In a world of increased communication of examination results mediated by information technology, there is a need for a common vocabulary...According to VIM in 2.1, note 1, "Measurement does not apply to nominal properties", so they cannot be a subject for metrology. However, most scientific disciplines, not only clinical laboratory sciences, also rely—some predominantly—on the description of properties without size.' A way forward is introduced in the next section, and culminates in Sect. 5.5, where a unified description is given of the response of a measurement system in terms of the concept of entropy which turns out to be a more robust descriptor than misclassification probabilities.

1.2.2 A Way Forward. Man As a Measurement Instrument

The two principal hallmarks of metrology, namely: traceability and measurement uncertainty, respectively (Chap. 3) are the two aspects—the exceptional homogeneity of the laws of physics which underpins quantification, and the heterogeneity which is more the essence of the social sciences—invoked by Nelson (2015) as fundamental distinctions between the sciences. With the explicit aim of developing a common language and complementary methodologies for cooperation about measurement and decision-making between sociologists, physicists and others, we have recently proposed (Pendrill 2010, 2014a, b) an approach that seems to be equally applicable to both physical and social measurements and combines both aspects.

Drawing simple analogies (Finkelstein 2009; McGrane 2015) between 'instruments' in the social sciences—questionnaires, ability tests, etc.—and engineering instruments such as thermometers does not go far enough. The analytical chemists have proffered a suggestion that 'examination' could replace 'measurement' in the case of nominal properties. They define the activity of examination as essentially consisting of 'comparing the property considered, i.e. the examinand (2.7) by way of an examining system (2.8), with the property of a "reference" of similar nature. Such a reference may be personal and subjective, such as a person's memory of a colour, or the reference may be objective, such as a nominal reference material (4.2).' (Nordin et al. 2018)

Our new approach (Pendrill 2014a, b, c, 2018) identifies instead the test person (e.g. patient) as the instrument at the centre of the measurement system. The key to our approach to describing measurements in the social sciences is to treat the human responder herself, rather than the questionnaire, as the instrument.

In Fig. 1.1 are shown two versions of measurement system where Man is being measured:

(a) Measurements with a conventional measurement system (Sect. 2.4) shown in Fig. 1.1a may be challenging owing to the complexity of a human being. Examples include measurement in healthcare, such as of body mass index or with a fever thermometer.
(b) A radically different viewpoint is to regard measurement systems where a human being acts as a measurement instrument at the heart of a measurement system (Fig. 1.1b).

The measurement system approach is essential in obtaining both aspects of metrology: traceability and uncertainty in all sciences:

- Metrological standards, reflecting fundamental symmetries, traceability to which can enable the comparability of measurements (and by extension even the intercomparability of products and services of all kinds), can only be established if one can separate out—with a process called restitution—the limiting factors of the measurement system used to make measurements from the response of the system to a given stimulus.

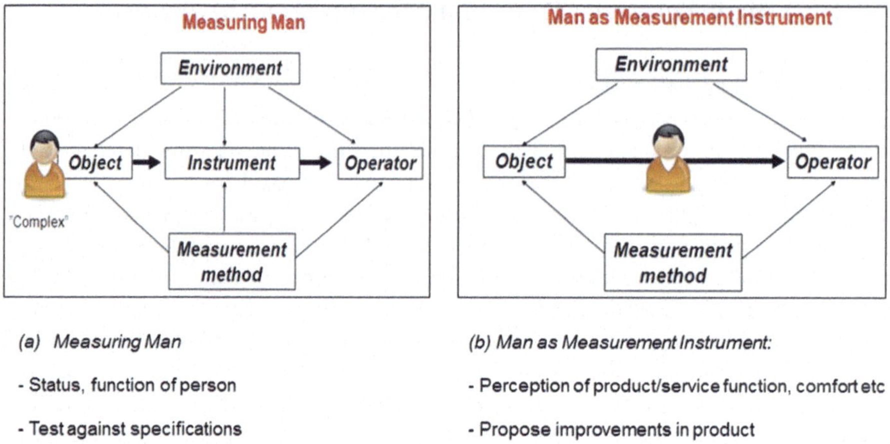

Fig. 1.1 Measuring Man (reproduced from Pendrill 2014b with permission)

- Less than complete information about a system leads to uncertainties, causing to incorrect decision-making, for example approval of an incorrect product. Formulation of a performance measure, i.e. how well a measurement system performs an assessment—actually measurement uncertainty—seems to be treatable with similar (categorical) methods, whether it is 'instruments'—questionnaires, examinations, etc.—in social science or in assessing how well a measuring instrument shows if an item or product is within or outside a specification limit.

1.2.3 Categorical Scales: Logistic Regression

A certain kind of measurement—on a categorical scale (Fig. 1.2)—turns out to be of particular interest for our purpose of dealing with quality-assured measurement across Social and Physical Sciences.

A categorical scale is one in which the results (responses of a measurement system) are sorted according to a (usually finite) number of discrete categories. In many cases, this is done for purely practical reasons, where one chooses to classify measurement responses in categories because time or other resources are limited and there is no need for a more complete, quantitative record of responses on a continuous scale. We regard such scales as the arena where measurement uncertainty coincides conceptually with indecision. Categorical scales are of course a primitive version which underlies more quantitative scales, but when the quality characteristics of objects (products, processes, phenomena, etc.) are less than fully quantitative, values can be arranged on a scale, such as the ordinal, where relations between distances among a discrete set of categories are not completely known,

Fig. 1.2 Categorical scale

apart from evidence for an approximate monotonic trend (Fig. 1.2). (A review of categorical data analysis can be found in the work of Agresti (2013)).

There is an increasing need across the disciplines to assure the quality of such categorical measurements in themselves, alongside assurance of more quantitative data. But the science of how to deal with measurement of categorical properties is as yet in its infancy. Categorical scales provide an arena for investigating the important conceptual interplay between measurement, uncertainty and decision-making and turns out to be a fertile ground for unifying measurement across the physical and social sciences.

Demands for quality-assured measurement on qualitative and categorical scales are increasing: Measures based on human observations are typically qualitative, of course in sectors where the human factor is obvious, such as healthcare; sensory science (Skedung et al. 2015); teaching (Wilson et al. 2015); citizens' understanding and information (Weller et al. 2013); services and safety and so on. The majority of questionnaires are based on categorical scales. Beyond these obvious human-factor sectors, many manufacturers of products of all kinds are becoming increasingly aware that customer perception and the 'soft reliability' of their products may be the key-determining factor (Pendrill 2010, 2014a, b, c; den Ouden et al. 2006, Sect. 1.1). The fraction non-conforming product, for example in manufacturing, obtained with attribute sampling is typically on a categorical scale (Pendrill 2008; Akkerhuis et al. 2017; Ahn 2014). Other qualitative measures include qualitative examination of images and patterns, e.g. in analytical chemistry (Hardcastle 1998; Ellison and Fearn 2005), forensics (Vosk 2015), healthcare (Mencattini and Mari 2015).

Our approach Man as a measurement instrument (Sect. 1.2.2) appears applicable to all these examples, from human-factor applications to qualitative material testing: In each case, a performance score is needed to characterise the interaction between a 'probe' and an 'item', in many diverse fields, be it the efficiency of care, diagnostic and software systems or materials testing, such as hardness. Sorting responses on a categorical scale is conceptually linked to ambiguity in decision-making associated with uncertainty.

Several quality characteristics of the object to be conformity assessed are thus likely to be categorical variables where the traditional methods of statistical and metrological characterisation are not immediately applicable (Svensson 2001). Alongside more quantitative measures of efficiency for instance, functional aspects such as satisfaction and effectiveness will often need special evaluation tools.

One set of tools is based on logistic regression (Tezza et al. 2011; Rice et al. 2018, Sect. 3.5), and includes the Rasch (1961) formulation of Generalised Linear Model (GLM)

$$S = z = \theta - \delta = \log\left(\frac{P_{\text{success}}}{1 - P_{\text{success}}}\right) \tag{1.1}$$

We interpret the Rasch model in terms of Man as a measurement instrument, where restitution of the stimulus S from the response P_{success} (eq. 2.8) yields a measure of instrument (person) ability, θ, expressed in the same unit (e.g. Lexiles 2018) as the object attribute, δ, such as task difficulty. In Table 1.1 are summarised some typical quality characteristics amenable with this approach.

Different scales of measurement—ranging from fully quantitative scales such as the ratio and interval which are commonly used in physics and engineering, to the more qualitative scales of the ordinal and nominal kind—will be described in Chap. 3. To date, many observations in healthcare, sustainability, material testing (such as hardness and rheological properties), etc., are considered to lie 'off the scale' of quantitative measurement. Fitting the Rasch model (Eq. (1.1)) establishes the scale of measurement, where the span of the scale is defined in part by the spread of person attribute (e.g. from the most able to the least) and in part by the spread of item attribute (e.g. from the most difficult to the easiest task). The first time this is done in a new application of the Rasch model, one is so to speak 'calibrating' the model (Chap. 4).

Table 1.1 Coupling item attributes to person characteristics in diverse responses for various applications (Pendrill 2014a, b)

Response	Item attribute (δ)	Person characteristic (θ)	Applications (examples)
Satisfaction	Quality of product	Customer leniency	Consumer electronics; Cosmetics; Health care; Services
Performance of task	Level of challenge of activity	(Dis-)ability	Citizen's understanding and information; Learning; Reading; Psychometrics; Rehabilitation
Accessibility (e.g. of transport mode)	Barrier hinder (or cost)	Utility (or net benefit, well-being, disability,...)	Commuter traffic; discrete choice and valued prospects

1.3 Opening the Quality Assurance Loop

The whole structure of the layout of this book about quality-assured measurement across Social and Physical Sciences will be based on the following essential steps for Conformity Assessment aiming to keep product on target:

(a) Define the entity and its quality characteristics to be assessed for conformity with specified requirements (Chap. 1).
(b) Set corresponding specifications on the measurement methods and their quality characteristics (such as maximum permissible uncertainty and minimum measurement capability) required by the entity assessment at hand (Chap. 1).
(c) Produce test results by performing measurements of the quality characteristics together with expressions of measurement uncertainty (Chaps. 2–5).
(d) Decide if test results indicate that the entity, as well as the measurements themselves, is within specified requirements or not (Chap. 6).
(e) Assess risks of incorrect decisions of conformity (Chap. 6).
(f) Final assessment of conformity of the entity to specified requirements in terms of impact (Chap. 6).

This basic structure has its roots in early efforts in industrial quality assurance during the twentieth century, where the steps can be found in the famous 'Plan, Do, Study, Act' (PDSA) cycle, emphasising that product quality is not only determined by actions at the point of production, but importantly at every step of the loop (Fig. 2.1)—from defining to delivering product—in what should be an on-going 'dialogue' between customer and supplier (Deming 1986). This builds on earlier ideas that manufacturing, according to Shewhart (1986), is a three-step process of specification, production and inspection. The same basic idea is still current in industry, but also in other fields such as healthcare, where a modern interpretation can be found in the structure, process and outcome model of Donabedian (1988). Deming's idea was that—if quality were depicted as a third dimension—then on each turn of the loop for the various processes of production, from initial product idea (step a), through manufacture to final product use and disposal (step f), quality would hopefully make a spiral of increasing value directed out of the loop, towards the reader. Appearing in early versions of the well-known ISO-9000 standards, the quality assurance loop or cycle continues to provide the overall structure of these widely used quality assurance standards. Initially intended for quality assurance in post-war product manufacturing, such standards have over the years found increasing application in the service industry and more recently in healthcare (EN 15224:2012). Whether you are in the broadest sense a 'producer' or a 'measurer', working in the social or physical sciences, we believe you will find following this quality assurance loop relevant [a,. . ., f; Fig. 2.1].

The reader is encouraged to formulate his or her own quality assurance loop, by choosing a particular product of interest. Starting in the present introductory chapter, a template will be provided (see final pages of this Chapter, for instance) which the reader is intended to fill in for the product of choice at successive stages in the quality assurance loop, so that by the end of the book, one should have a complete, worked-out evaluation of quality assurance for that product.

1.4 Specification of Measurement Problem: Entity Specifications

The first steps in any measurement task as specified in the quality assurance loop (Sect. 1.3) are to identify the entity of interest and set specifications on its characteristics. An example of the first step—describing the product—will be given in the case study of Sect. 1.5. We choose a case of pre-packaged goods—more specifically coffee powder—as a simple example followed throughout this book, which will serve to illustrate measurement across the physical and social sciences.

1.4.1 Entity Specifications

When keeping product 'on target', key Conformity Assessment concepts such as 'entity' and 'quality characteristic' are those defined for instance in ISO 10576-1:2003. The wide range of entities and their characteristics which may be of interest is illustrated in Table 1.2. This can encompass everything from a single physical object, quantitatively characterised, to a service scored qualitatively, as needed when attempting a unified picture of quality-assured measurement across the Social and Physical Sciences.

A major link between the present introductory chapter—with its focus on entity (or object) and its characteristics—and the rest of the book dealing with measurement is through the observation that a main element of a measurement system (to be described in Chap. 2) is indeed the *measurement object*. The measurement object is an inseparable part of the measurement system, and of course the principal aim of any measurement is to estimate the quantity of the property of interest associated with the entity.

Irrespective of whether we are considering quantitative or qualitative observations anywhere in the physical or social sciences, production and measurement

Table 1.2 Entities and characteristics (ISO 10576-1)

Entity	Characteristic
A weight	• Mass
A lot of sugar bags	• Average mass per bag • Homogeneity (standard deviation of mass of bags) • Percentage of bags with conforming masses
Treatment of a specific disease	• Average number of cured patients • Average waiting time for treatment

dispersion are often confounded and there is a long-standing lack of clarity in concepts, definitions and nomenclature which arises at the interface where two disciplines—Metrology and Conformity Assessment—meet. Two principally distinct, but closely related and easily confounded concepts coming from the two disciplines are, respectively:

- A 'quality characteristic': a quantity intended to be assessed, the subject of this chapter.
- A 'measurand': a quantity intended to be measured, dealt with in Chap. 2.

A few, clear illustrations of this dichotomy can be found in the literature, such as exemplified in the concept diagram: Figure B.16 of ISO 3534-2:2006 Applied Statistics, where the distinct pairs 'quantity: measurement' and 'characteristic: test' are juxtaposed (Fig. 1.3).

The 'construct', a property of the measurement object, e.g. the difficulty of a task or the quality of a service, is distinct from other properties of the measurement system, such as sensitivity and ability, when 'probing' the construct, to be described later in this book.

With a focus on the object or entity, rather than the measurement, it is important to specify the assessment target as clearly as possible (Rossi and Crenna 2016):

- 'Global' conformity denotes the assessment of populations of typical entities.
- 'Specific' conformity assessment refers to inspection of single items or individuals.

In evaluating *product* variations of the quality characteristic $\eta = Z$ in the 'entity (or product) space',[2] measurements might be made on repeated items in a production process or by taking a sample of the population of items subject to conformity assessment (ISO/IEC 17000:2004; Pendrill 2014c). Examples of these 'intrinsic' entity variations include differences in manufacture or when an entity is subject to the wear and tear of use. The corresponding probability density function (PDF), $g_{\text{entity}}(z)$, of the product characteristic will have a form determined ideally (in the absence of measurement or sampling uncertainty) by the variations in the inherent characteristic of product, process or system of prime interest in conformity assessment.

The established discipline of statistical quality control, including hypothesis testing on process parameters (with point and continuous estimators), is described extensively in the literature, for instance by Montgomery (1996). The measurand—in the terminology of conformity assessment, a 'quality characteristic' of the entity intended to be assessed (Fig. 1.3)—may be, as in statistics, either a measure of:

[2]We will use a separate notation for the measurement space—see Chap. 2

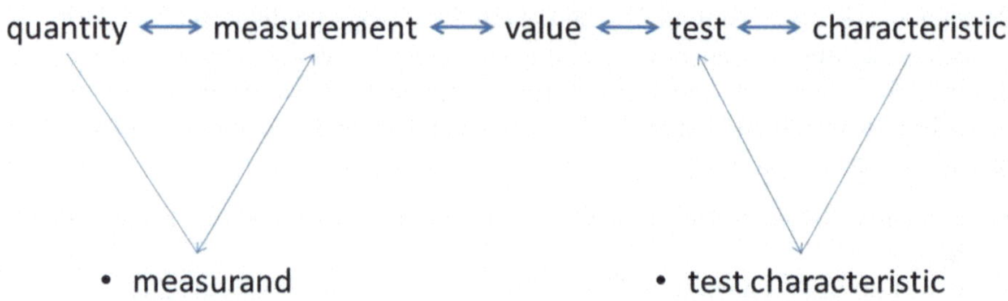

Fig. 1.3 Concept diagram: Determination of characteristics and quantities (adapted from Figure B.16 of ISO 3534-2:2006)

- 'location', e.g. a direct quantity of the entity, such as the mass of a single object, an error in mass (deviation from a nominal value), or an average mass of a batch of objects,
- 'dispersion', e.g. the standard deviation in mass among a batch of objects (Table 1.2).

As will be discussed in more depth later, models of measurement will need to be able to deal with:

- The actual result of measurement, that is, estimating the measurand or quality characteristic of the entity of interest,
- Propagation of measurement bias through the measurement system,
- Propagation of variances through the measurement system.

1.4.2 Definition of Test Problem, Based on Product Description

When making conformity assessment decisions aimed at keeping product 'on target' (Sect. 1.1), descriptive statistics is used when setting prescriptive requirements (tolerances) on a quality characteristic, z, of a product in terms of:

- Specification limits, U_{SL} and L_{SL}, for the magnitude of a characteristic of any entity,
- For any entity, the maximum permissible (entity) error, *MPE*.

For a symmetric, two-sided tolerance interval: $MPE = (U_{SL} - L_{SL})/2$.

For a one-sided interval: $MPE = U_{SL}$–nominal, for instance.

A process capability index, C_p, can be defined, for instance, in terms of estimated product variations as: $C_p = \frac{U_{SL} - L_{SL}}{N \cdot s_p}$ with standard deviation s_p and where $C_p = 2$ is used the famous 'six-sigma' approach to statistical process control (SPC) (Joglekar 2003).

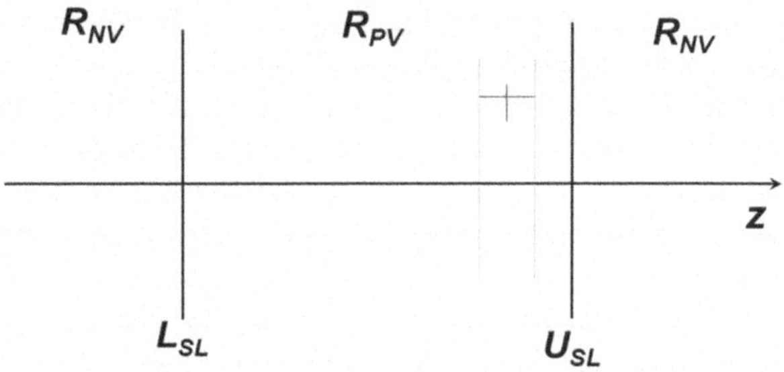

Fig. 1.4 Example of division of domain for a characteristic, z, [adapted from ISO 10576-1], together with an uncertainty interval

Typical decision rules in conformity assessment are of the form: 'Conformity to requirements is assured if, and only if, the uncertainty interval of the measurement result is inside the region of permissible values R_{PV}' [ISO 10576-1], where that region is bounded, as shown in Fig. 1.4, above and below specification limits, by regions of non-permissible values, R_{NV}.

Note that entity specifications are normally set on the basis not only of what can be practically and economically produced, but also ultimately on what the consumer or other end-user requires in terms of product characteristics. Meeting these, sometimes contradictory, demands leads to a number of major challenges typically encountered when handling the quality characteristics of entities subject to conformity assessment, perhaps particularly so in the social sciences:

- There is a plethora of potential quality characteristics of the entity of interest to the 'end-user', implying the need for structured approaches to describing and prioritising the constructs, including both functional and non-functional requirements (Sect. 5.2.3).
- A 'design of product' process, analogous to design of experiment in statistics and measurement, needs to be made prior to production, especially where risks and benefits are to be balanced proactively.

These will be exemplified for each of the various aims of conformity assessment: consumer confidence and producer and supplier tools for quality assurance for a wide variety of reasons (Sect. 1.1).

1.4.3 Structural Models and Specifications of Entity Characteristics

Having identified the entity and some of its quality characteristics, the next step in the quality loop (Sect. 1.3 step (a); Fig. 2.1) is to make a structural model which, with various approaches, is used to formulate relations between different quantities characteristic of the quality of the entity. The structural model will be used to summarise, for both descriptive and prescriptive purposes, our knowledge and understanding of which factors determine product quality. In some cases the structural model will enable prediction of future values of the product, for instance when planning production and designing. This is, of course, a vast subject in itself. But again: while this is a book primarily about measurement, we can permit ourselves a brief and cursory description of the production process in this introductory chapter, both in terms of motivating measurement as well as drawing analogies where measurement processes are regarded as a particular kind of production process.

Typical descriptions of constructs, for instance in the social sciences, can be found in quotes from studies of customer satisfaction: 'Some variables cannot be observed directly. Examples of such are intelligence, depression, suffering, attitudes, opinions, knowledge of something, satisfaction. Analysis of these variables can only be performed indirectly by employing proxy variables. The former (unobserved variables) are referred to as latent variables, whilst the latter (proxy variables) are known as observed variables' (de Battisti et al. 2010). Further examples of construct description will be given below and the distinction between functional and non-functional characteristics is made in the next section (Sect. 1.4.4). A case study of the perception of the 'prestige' experienced with pre-packaged coffee will be evaluated in Sect. 4.5.2.

A general, construct specification equation can be formulated:

$$z = f[x_1, \ldots, x_m] \tag{1.2}$$

which involves sorting all variables into two groups—the dependent variable, z, on the LHS of Eq. (1.2), and a number of independent variables, x, on the RHS. A certain expression of z as a function of x (Eq. (1.2)) describes how values of the entity construct z—a 'response'—are related to a set of 'explanatory' variables x. (The particular case where the entity responding is a measurement system will be the main focus of most of this book, from Chap. 2 on.) Modelling encompasses firstly setting up the expression, based on what is known about the system or entity of interest. A common visualisation of explaining product in terms of 'cause and effect' can be made by drawing an Ishikawa ('fishbone') diagram, exemplified in Fig. 1.5, where a series of 'bones' (one for each independent variable, x) converges to produce the overall response, z, which depends on them.

Design of experiments in traditional statistics means the process of systematically varying controllable input factors to a 'manufacturing' process (in the broadest sense) so as to demonstrate the effects of these factors on the output of production (Montgomery 1996) and is one important application where Eq. (1.2) and the Ishikawa diagram (Fig. 1.4) come into play, not only in manufacturing but also more broadly throughout the physical and social sciences. Once the causes and effects are known and 'explained', then remedies for imperfections can be considered. This essentially active method of design of experiment (DoE) can be contrasted with the more passive statistical process control (SPC). Quoting Montgomery (1996): 'Statistically designed experiments are invaluable in reducing the variability in the quality characteristics and in determining the levels of controllable variables that optimize process performance'.

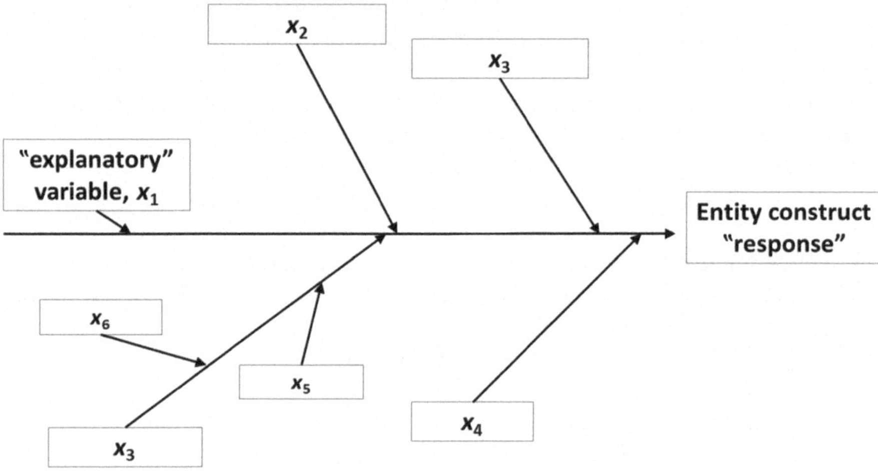

Fig. 1.5 Ishikawa diagram visualising 'cause and effect' in construct specification (Eq. (1.2))

The multivariate relations developed here (Eq. (1.2) and following) will enable prediction of the properties of an arbitrary object, as a kind of 'recipe', based on previous experimental investigations. A manufacturer can use such construct specification equations in order to 'tailor-make' a product. A clinician can design new cognitive tests for Alzheimer patient once a construct specification equation for task difficulty has been formulated (Stenner and Smith 1982, Sect. 5.2.2). In the context of conformity assessment, specifications will be set prescriptively on characteristics which appear on either side of the entity construct specification Eq. (1.2).

A useful illustration of formulating Eq. (1.2) is in terms of multi-attribute alternatives (Roberts 1985, pp. 197ff) as multidimensional explanatory variables. A set of alternatives can be identified:

$$\mathbf{a} = (a_1, a_2, \ldots, a_n) \tag{1.3}$$

where a_i = rating of alternative **a** on the *i*th attribute (dimension). One example from healthcare is where

$$\text{Care efficiency} = f(\text{expenditure, bedtime, complications}\ldots)$$

The various independent attributes in this expression can be of quite different kinds:

a_1 = amount of money spent on treatment, drugs, etc.
a_2 = number of days in bed with high index of discomfort
a_3 = number of days in bed with medium index of discomfort
a_4 = number of days in bed with low index of discomfort
$a_5 = 1$, complication A occurs; $a_5 = 0$, complication A does not occur
$a_6 = 1$, complication B occurs; $a_6 = 0$, complication B does not occur

$a_7 = 1$, complication C occurs; $a_7 = 0$, complication C does not occur

and so on, that is, including categorical measurements of the kind shown in Fig. 1.1.

Decisions about care can be made on the basis of quantitative comparisons, for instance of the 'indifference' *E*, between different care forms:

$$(a_1, a_2, a_3, a_4, a_5, a_6, a_7)E(a_1, 0, a_3', a_4, a_5, a_6, a_7)$$

This expression indicates that $a_2 = 0$; zero days in bed with high index of discomfort can be 'traded' against and is considered equivalent to $a_3' =$ number of days in bed with medium index of discomfort.

Other examples (Roberts 1985, p. 197ff) include:

$$\text{Choice of job} = f(\text{salary, job security, advancement possibilities, } \ldots)$$
$$\text{Usability} = f(\text{efficiency, effectiveness, satisfaction} \ldots)$$

Sections 5.2.2 and 5.4.2 explore similar formulations found in multivariate studies of chemometrics and in the 'House of Quality' approach.

As elsewhere in this book, measurement will be considered a special type of production, so one can expect to be able to formulate a construct specification Eq. (1.2) even for measurement. Roberts case of multi-attribute alternatives (Eq. (1.3)), Mental testing = *f(individuals, test items…)*, is such a special case where the structural model applies to a measurement system. Our approach is to single out measurement as the specific case of a construct specification Eq. (1.2) where it is the 'response' of a measurement system which is of interest, as is developed further in Chap. 2 on.

1.4.4 Construct System and Structural Modelling: Functional and Non-functional Characteristics

Understanding of the very general construct specification Eq. (1.2) can be enhanced with visualisations of which entity has the constructs of interest and how these are related. A useful picture exemplified in Fig. 1.6 is of the interplay between three main actors, presented as a triangle of interactions between a user who uses a tool to perform a task.

In the example shown in Fig. 1.6, the overall concept 'ergonomy' (Kuijt-Evers 2006) is a result of a sum on interactions on all sides of the triangle between three elements—user, tool and task. Each element is characterised in terms of an attribute—such as user ability or task difficulty. A user's ability will be assessed by how well she performs the task at hand—e.g. driving in a nail—which has a certain level of difficulty. If the user employs a hammer for the task, then overall satisfaction with the tool will be determined not only in terms of the quality of the tool but also the

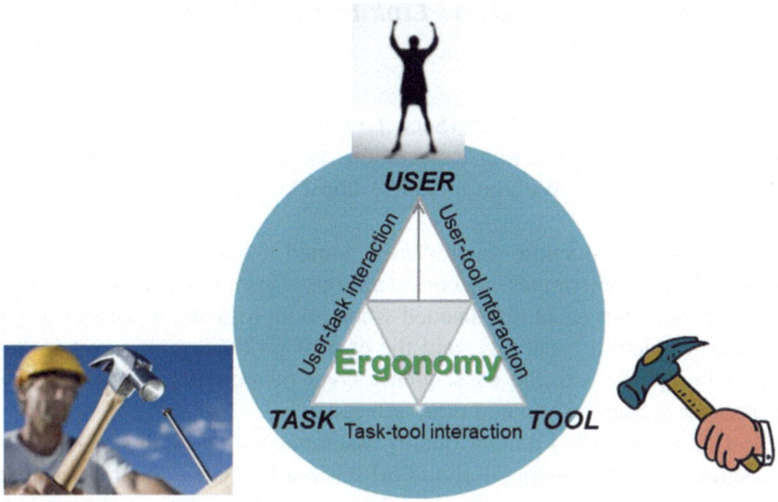

Fig. 1.6 System of interplay between a user who uses a tool to perform a task (adapted from Kuijt-Evers 2006)

leniency of the user. The different explanatory variables, x, sum together to produce an overall construct, e.g. ergonomy, as the dependent variable, the response z, as expressed by the construct specification Eq. (1.2). This approach can be extended to descriptions of other constructs characterising different entities (Dagman et al. 2013). Section 5.2.3 develops a general procedure for modelling construct specification equations.

In the corresponding case of the construct 'usability', measures of (a) efficiency ('doing things right' in a non-functional sense) might be made in terms of the time taken to perform a task, complemented with measures of (b) satisfaction—where patients are asked in surveys how pleased they are with the health care services provided, and (c) effectiveness ('doing the right things' in a functional sense), that is, actually improving a patient's health; with an overall quality characteristic termed 'usability' (ISO 9241 1998, 2018; ISO/TS 20282-2:2013).

Measurements made on each side of the triangle shown in Fig. 1.6 may be either:

- quantitative information will be associated with: (A) Test of *non-functional* characteristics of product; is the product 'correctly' made?
- or qualitative information: (B) Test of product *function*; is the product 'right'?

Indeed, what is considered 'functional' has in fact changed meaning in recent years. The traditional focus on 'correctly' made products from the technological point of view has in the past decade or so shifted to considerations of whether the product actually functions 'right' for the end-user (Sect. 1.1).

1.4.5 Uncertainty and Risks: Link to Final Steps in Quality Loop

If 'quality' is meeting expectations (Sect. 1.1), then 'uncertainty' represents limitations on quality, i.e. not meeting expectations because of lack of knowledge and control about the object. Without complete knowledge, there are risks that things will 'go wrong'.

When planning production, much can be gained if one can proactively plan for 'fitness for purpose', encompassing considerations—prior to production—of aspects such as the production capability needed when about to make a product or process of a certain production capability, and the cost of production at a certain level of uncertainty compared with the consequence costs of incorrect decisions of conformity, customer dissatisfaction and other effects.

The last steps ((e) and (f)) in the quality assurance loop described in (Sect. 1.3) in any measurement task, which will round off this book in Chap. 6, are to account for actual decision risks before finally assessing conformity of the entity to specified requirements in terms of impact.

There is increased emphasis on risk management in the latest version of the product quality assurance written standard ISO9001:2015, which states:

> **0.3.3 Risk-Based Thinking** Risk-based thinking ... is essential for achieving an effective quality management system. The concept of risk-based thinking has been implicit in previous editions of this International Standard including, for example, carrying out preventive action to eliminate potential nonconformities, analysing any nonconformities that do occur, and taking action to prevent recurrence that is appropriate for the effects of the nonconformity...To conform to the requirements of this International Standard (ISO 9001:2015), an organization needs to plan and implement actions to address risks and opportunities. Addressing both risks and opportunities establishes a basis for increasing the effectiveness of the quality management system, achieving improved results and preventing negative effects.
>
> Opportunities can arise as a result of a situation favourable to achieving an intended result, for example, a set of circumstances that allow the organization to attract customers, develop new products and services, reduce waste or improve productivity. Actions to address opportunities can also include consideration of associated risks. Risk is the effect of uncertainty and any such uncertainty can have positive or negative effects. A positive deviation arising from a risk can provide an opportunity, but not all positive effects of risk result in opportunities. (ISO9001:2015).

Risk will not only be a measure of the probability, p, of a risky situation, but will also include a measure of the impact, C, of event in case it occurs: that is, risk $= p \cdot C$.

The next Chap. 2 will deal with amongst others design of experiment, where measurement effort is ideally decided upon proactively, by balancing the costs of measurement against the consequence costs of incorrect decisions of conformity.

Measurement uncertainty is one example of uncertainty, but not all uncertainties arise because of limited measurement quality. As explained, 'uncertainty' represents limitations on quality, i.e. not meeting expectations because of lack of knowledge and control about the object. In explaining the motive for the expression of risk in the context of uncertainty about product (e.g. electrical energy in a smart grid), Yang and Qiu (2005) note: 'The higher the uncertainty is, the higher the riskiness; the higher the expected utility of an action, the less the riskiness'. We will return to this discussion in Chap. 6 (e.g. Eq. (6.6)).

1.5 Case Study: Fit-for-Purpose Measurement Specification

This introductory chapter concludes with a simple example, about *pre-packaged coffee*, intended to exemplify the principal contents of the chapter in the case where the entity is an example of a generic product, and how to use the templates at the end of the chapter. A wide variety of pre-packaged products (e.g. food, cosmetics, etc.) have a major role to play in national economies and in trade, and are therefore subject to legal metrological regulation where goods are labelled in predetermined constant nominal quantities of mass, volume, linear measure, area or count (OIML 2016). Legal metrology will be cited at several places in this book since it provides a number of important examples with which the methods of optimised measurement uncertainty and cost operating characteristics (Pendrill 2014c, Chaps. 2 and 6) can be demonstrated.

1.5.1 Describe the Product [§E1.1]

As part of the process of describing the entity construct and establishing grounds for prioritisation, as described in Sect. 1.4, typical responses based on the viewpoints of both producer and consumer of pre-packed coffee can be entered in the template at the end of this chapter [§E1.1]:

What is the product used for?	*Produce a cup of coffee from powder*

Which properties of the product are important?	• *The packet should contain as promised - the correct amount powder, good quality etc.* • *The packet should keep the powder freshness longer* • *Easy to open and re-seal the packet* • *One experiences a certain 'prestige' when serving this brand of coffee*

These are some of the properties which would enter into the construct specification Eq. (1.2) and Ishikawa Fig. 1.5, perhaps as multi-attribute alternatives (Eq. (1.3)). A useful exercise would be to formulate a system picture according to the system picture of Fig. 1.6 where, instead of hammering nails, the consumer would interact both with each packet of coffee (a 'tool') as well as different tasks of everyday living where coffee-drinking can be important.

Are certain functions or properties of the product regulated because of safety or similar reasons?	Yes, mass content in packet is regulated wrt consument protection through legal metrology. 500 g coffee packet should contain at least 485 g

With a view to considering pro-actively design of product (Sect. 1.4), at this early stage in formulating a measurement task it will be useful to give estimates of various costs, including the impact if things go wrong.

What will be the consequences if the product doesn't work correctly?	The consumer will not get good coffee and will become dissatisfied. She may even choose another coffee mark in the future and tell her friends
Can a value be assigned to the consequence costs?	If the consumer stops buying this brand of coffee, losses may be 120€/year per consumer

1.5.2 Product Demands [§E1.2]

Tolerances can be set (Sect. 1.4.2) for a number of multi-attribute (dimensional) alternatives (Eq. (1.3)) readily identified for the pre-packaged goods, where quality characteristics can be either 'functional' or 'non-functional' (Sect. 1.4.4).

What are the 'optimal' values of the product's most important characteristics?	The packet should contain as promised: - correct amount powder - 500 g - good quality ○ taste (roasting 100%[1], body 95%, acidity 5% according to manufacturer's data sheet (Bastin 2015))
	The packet should keep the powder freshness longer: • storage time 6 months
	Easy to open and re-seal the packet: • Take less than 1 min to open packet Resealable so powder freshness is maintained according to above specs
	One experiences a certain 'prestige' when serving this brand of coffee: Measure of perceived 'prestige': 80%

[1] On a scale rated 0 – 100%

Part of risk assessment (Sect. 1.4.5) which will also be referred to during design of measurement:

How much will your costs vary with varying deviations in product characteristics?	*The packet should contain as promised:* - *correct amount powder - at least 485 g => (10€/500 g)*15 g = 0.3€ per packet = 3.6€/year*
	- *good quality:* *taste (roasting 90%, body 90%, at least acidity 3% and at most 6%) => customer dissatisfaction: loss maybe 120€/year*

1.5.3 Definition of Test Problem, Based on Product Description §E1.1

Four different kinds of entity (product) tests can be usefully identified:

A Test of *non-functional* characteristics of product; is the product 'correctly' made (Sect. 1.4.4)?

B Test of product *function*; is the product 'right' (Sect. 1.4.4)?

C Initial verification of product,

D Subsequent verification of product.

Each of these different kinds of product test will be exemplified below for the specific example of pre-packaged coffee. Using the template at the end of this chapter (which the reader is encouraged to fill in for the particular case chosen), some typical responses could be:

Consumer requirements aimed at ensuring a guarantee that each coffee packet is not sold at underweight obliges the producer to test that manufactured coffee packets of, say, nominal mass 500 g weigh at least 485 g. This one-sided, lower specification limit (illustrated in Fig. 1.4) on the quality characteristic, z, the mass per packet, has a corresponding *MPE* of—15 g, as stipulated (in the simplest case) in current EU legislation for pre-packaged goods. Typically, a coffee producer might fill about 100,000 packets à 0.5 kg a day. Commodity value is typically 10 €/kg when on the market, while production costs will be some fraction so that the producer can make a profit. These economic factors can be used to optimise production, by balancing these against the consequence costs associated with dissatisfied customers.

A Test of non-functional characteristics of product/system. Is the product 'correctly' made?

Describe scope of product test (entity, characteristic, test range)	*Protective film between powder and packaging, liquid penetrability [2], 0 – 100%*

Further work could be done for example to find an expression which relates the humidity (response) of the coffee powder to the penetrability (explanatory variable) of the protective film.

B Test of product/system function; is the product right?

Describe scope of product test (entity, characteristic, test range)	*(a) The packet of coffee powder should contain as promised:* - *Mass per packet: 300 g - 700 g* - *powder of 'good' quality* o *taste (roasting 60 – 100%[3], body 60 – 100%, acidity 0 – 10%)*
	(b) Easy to open and re-seal the packet: • *Time between 1 s and 5 min* *Packet, powder freshness after opening, 1 month - 1 year*

[1] On a scale rated 0 – 100%

Further work could be done for example to find a construct specification Eq. (1.2) which relates the functional durability of the coffee powder in terms of freshness (response) to the humidity (explanatory variable), which in turn was considered a non-functional characteristic related to film properties considered under test A. Several of these functional characteristics (taste, ease) will have to be suitably transformed based on the Rasch GLM expression (1.1) to reflect their ordinal character. See further in Chap. 3 and Sect. 4.3.2.

C Initial verification of product (Fig. 1.7)

Describe the product test:	*Non-functional test of protective film as above to be made on 'prototype' packet*
- Choose verifications module (A–H)	- *Verification module F* - see Fig 1.7*

*Conformity assessment module F stipulates that 'a notified body…shall carry out appropriate examinations and tests…' (EU commission 2019)

D Subsequent Verification

Describe the test of product:	*Sampling of packet contents by weight made during production runs by manufacturer*
	- *Verification module D* - see Fig 1.7*

*Conformity assessment module D stipulates that the 'Manufacturer of the product shall operate an approved quality system for…final product inspection and testing…' (EU commission 2019)

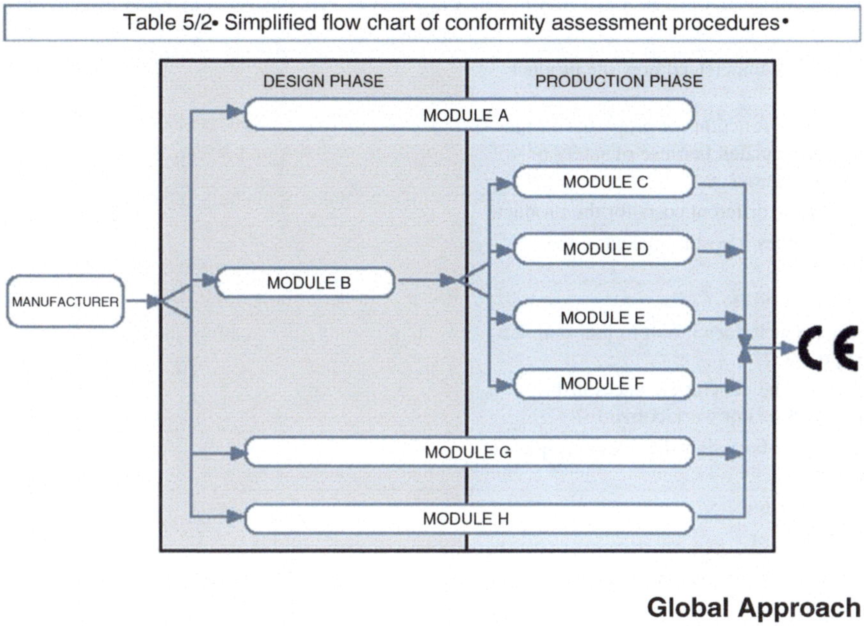

Fig. 1.7 Modules of conformity assessment according to the European 'global approach' for products subject to CE-mark certification (EU commission 2019)

Exercise 1: The Product

E1.1 Describe the Product

	Your answers. . . . .
Please attach a specification sheet or similar about the product	
What is the product used for?	
Is the product part of a composite product?	
Are you the producer, supplier or user of the product?	
What is essential when choosing a certain product?	
Why is this product so much 'better' than the others?	
Is it better because of product development and/or increased demands from a particular application?	
Which properties of the product are important?	
What are the more important functions of the product?	

(continued)

	Your answers. . . . .
Is it easy to understand how the product works?	
Are certain functions or properties of the product regulated because of safety or similar reasons?	
Specify the different costs for the product:	
What does it cost to make/use the product?	
Product price	
How much does it cost to use/maintain the product?	
What will be the consequences if the product does not work correctly?	
Can a value be assigned to the consequence costs?	
Other questions:	

E1.2 Product Demands

	Your answers. . . . .
What are the 'optimal' values of the product's most important characteristics?	
How large deviations from these optimum values can be tolerated?	
How much will your costs vary with varying deviations in product characteristics?	
Other demands:	

E1.3 Definition of Test of Product, Based on Product Description §E1.1

	Your answers. . . . .
A. Test of non-functional characteristics of product/system. Is the product 'correctly' made?—Describe the test	

(continued)

	Your answers. . . . .
– Describe scope of product test (entity, characteristic, test range)	
– Describe environmental tests on product	
– How much does each product test cost?	
B. Test of product/system function. Is the product right—Describe the test:	
– Describe scope of product test (entity, characteristic, test range)	
– Describe test of 'hard' and/or 'soft' functions of product	
– Describe environmental tests on product	
– Describe possible safety tests of product	
– How much does each product test cost?	
C. Initial verification of product—Describe the test:	
– Choose verifications module (A–H)	
D. Subsequent verification of product—Describe the test:	
Others:	

References

A. Agresti, *Categorical Data Analysis*, 3rd edn. (Wiley, Hoboken, 2013). ISBN 978-0-470-46363-5

H. Ahn, Effect modeling of count data using logistic regression with qualitative predictors. Engineering **6**, 758–772 (2014). https://doi.org/10.4236/eng.2014.612074

T. Akkerhuis, J. de Mast, T. Erdmann, The statistical evaluation of binary test without gold standard: robustness of latent variable approaches. Measurement **95**, 473–479 (2017). https://doi.org/10.1016/j.measurement.2016.10.043

J. Dagman, R. Emardson, S. Kanerva, L.R. Pendrill, A. Farbrot, S. Abbas, A. Nihlstrand, Defining comfort for heavily-incontinent patients assisted by absorbent products in several contexts, in *Innovating for Continence Meeting*, Chicago, IL, 12–13 April 2013, (2013)

F. de Battisti, G. Nicolini, S. Salini, The Rasch Model in customer satisfaction survey data. Qual. Technol. Quant. Manag. **7**, 15–13 (2010)

W.E. Deming, *Out of the Crisis* (Massachusetts Institute of Technology, Center for Advanced Engineering Study, Cambridge, MA, 1986). ISBN 0911379010. OCLC 13126265

E. den Ouden, Y. Lu, P.J.M. Sonnemans, A.C. Brombacher, Quality and reliability problems from a consumer perspective: an increasing problem overlooked by businesses? Qual. Reliab. Eng. Int. **22**, 821–838 (2006)

A. Donabedian, The quality of care: how can it be assessed? JAMA. **260**(12), 1743–1748 (1988). https://doi.org/10.1001/jama.1988.03410120089033
S. Ellison, T. Fearn, Characterising the performance of qualitative analytical methods: statistics and terminology. TRAC-Trend Anal. Chem. **24**(6), 468–476 (2005)
EN 15224:2012 *Health Care Services—Quality Management Systems—Requirements Based on EN ISO 9001:2008*
EU Commission 2019 *CE Marking*, https://ec.europa.eu/growth/single-market/ce-marking_en
R. Feynman, *The Feynman Lectures on Physics*, Volume I (2013), http://www.feynmanlectures.caltech.edu/I_01.html#Ch1-S1
L. Finkelstein, Widely-defined measurement – an analysis of challenges. Measurement **42**, 1270–1277 (2009)
W. Hardcastle, *Qualitative Analysis: A Guide to Best Practice* (Royal Society of Chemistry, Cambridge, 1998)
ISO 10576-1 "Statistical Methods – Guidelines for the Evaluation of Conformity with Specified Requirements", 2003
ISO 2018, *What is Conformity Assessment?* https://www.iso.org/conformity-assessment.html
ISO 3534-2:2006, Statistics - Vocabulary and Symbols - Part 2: Applied statistics, International Organization for Standardization, Geneva
ISO 9001:2015 *Quality Management Systems — Requirements*, https://www.iso.org/obp/ui/#iso:std:iso:9001:ed-5:v1:en
ISO 9241-11:1998. Ergonomic Requirements for Office Work with Visual Display Terminals. Website. www.iso.org/iso/catalogue_detail.htm?csnumber=16883 Accessed 18 May 2017
ISO 9241-11:2018. Ergonomics of Human-System Interaction - Part 11: Usability: Definitions and Concepts, https://www.iso.org/standard/63500.html
ISO/IEC 17000:2004, *Conformity Assessment – General Vocabulary*, International Organization for Standardization, Geneva
ISO/TS 20282-2:2013. IDT Usability of Consumer Products for Public Use – Part 2: Summative Test Method. Website. www.iso.org/iso/catalogue_detail.htm?csnumber=62733 Accessed 18 May 2017
JCGM 200:2012 International vocabulary of metrology—basic and general concepts and associated terms (VIM 3rd edition) (JCGM 200:2008 with minor corrections) *WG2 Joint Committee on Guides in Metrology (JCGM)* (Sevrès: BIPM)
A.M. Joglekar, *Statistical Methods for Six Sigma in R&D and Manufacturing* (Wiley, Hoboken, 2003). ISBN: 0-471-20342-4
L. F. M. Kuijt-Evers, *Comfort in Using Hand Tools: Theory, Design and Evaluation* (2006), ISBN-10: 90-5986-218-X, ISBN-13: 978-90-5986-218-X
Lexiles, *Lexile Framework for Reading* (2018), https://lexile.com/
L. Mari, A. Maul, D.T. Irribarra, M. Wilson, Quantities, quantification, and the necessary and sufficient conditions for measurement. Measurement **100**, 115–121 (2017). https://doi.org/10.1016/j.measurement.2016.12.050
J. McGrane, Stevens' forgotten crossroads: the divergent measurement traditions in the physical and psychological sciences from the mid-twentieth century. Front. Psychol. Hypothesis Theory **6**, 1–8 (2015). https://doi.org/10.3389/fpsyg.2015.00431. art. 431
A. Mencattini, L. Mari, A conceptual framework for concept definition in measurement: the case of 'sensitivity'. Measurement **72**, 77–87 (2015)
D.C. Montgomery, *Introduction to Statistical Quality Control* (Wiley, Hoboken, 1996)
R. Nelson, Physics envy: get over it. Iss. Sci. Technol. **31**, 71–78 (2015)
G. Nilsson, Private Communication (1995)
G. Nordin, R. Dybkaer, U. Forsum, X. Fuentes-Arderiu, F. Pontet, Vocabulary on nominal property, examination, and related concepts for clinical laboratory sciences (IFCC-IUPAC Recommendations 2017). Pure Appl. Chem. **90**(5), 913–935 (2018). https://doi.org/10.1515/pac-2011-0613

OIML, "Quantity of product in prepackages", *International Recommendation* **OIML R 87** Edition 2016 (E), (2016), https://www.oiml.org/en/files/pdf_r/r087-e16.pdf

L.R. Pendrill, Assuring measurement quality in person-centred healthcare. Meas. Sci. Technol **29**(3), 034003 (2018). https://doi.org/10.1088/1361-6501/aa9cd2

L.R. Pendrill, Operating 'cost' characteristics in sampling by variable and attribute. Accred. Qual. Assur. **13**, 619–631 (2008)

L.R. Pendrill, Risk assessment and decision-making risk assessment and decision-making, in *Theory and Methods of Measurements with Persons*, ed. by B. Berglund, G. B. Rossi, J. Townsend, L. R. Pendrill, (Psychology Press, Taylor & Francis, Abingdon, 2010). ISBN: 978-1-84872-939-1

L.R. Pendrill, El ser humano como instrument de medida. *e-medida* (2014a)

L.R. Pendrill, Man as a measurement instrument. NCSLI Meas. J. Meas. Sci. **9**, 24–35 (2014b)

L.R. Pendrill, Using measurement uncertainty in decision-making & conformity assessment. Metrologia **51**, S206 (2014c)

G. Rasch, On general laws and the meaning of measurement in psychology, in *Proceedings of the Fourth Berkeley Symposium on Mathematical Statistics and Probability*, (University of California Press, Berkeley, 1961), pp. 321–334

S. Rice, L.R. Pendrill, N. Petersson, J. Nordlinder, A. Farbrot, Rationale and design of a novel method to assess the usability of body-worn absorbent incontinence care products by caregivers. J. Wound Ostomy Continence Nurs. **45**, 456–464 (2018). https://doi.org/10.1097/WON.0000000000000462. open access

F.S. Roberts, Measurement theory with applications to decision-making, utility, and the social sciences, in *Encyclopedia of Mathematics and its Applications*, vol. 7, (Cambridge University Press, Cambridge, 1985). ISBN 978-0-521-30227-2

G.B. Rossi, F. Crenna, "Toward a formal theory of the measuring system", IMEKO 2016 TC1-TC7-TC13. J. Phys. Conference Series **772**, 012010 (2016). https://doi.org/10.1088/1742-6596/772/1/012010

K. Sawyer, H. Sankey, R. Lombardo, Over-measurement. Measurement **93**, 379–384 (2016). https://doi.org/10.1016/j.measurement.2016.07.034

W.A. Shewhart, *[1939]. Statistical Method from the Viewpoint of Quality Control* (Dover, New York, 1986). ISBN 0486652327. OCLC 13822053. Reprint. Originally published: Washington, DC: Graduate School of the Department of Agriculture, 1939

L Skedung, L.R. Pendrill et al., Evaluation of sensory properties of skin creams using Rasch transformation, in *11th Pangborn Sensory Science Symposium*, Göteborg, 2015

J. Stenner, M. Smith, Testing construct theories. Percept. Mot. Skills **55**, 415–426 (1982). https://doi.org/10.2466/pms.1982.55.2.415

E. Svensson, Guidelines to statistical evaluation of data from rating scales and questionnaires. J. Rehabil. Med. **33**, 47–48 (2001)

wD.A. Swyt, *Challenges to NIST in Dimensional Metrology: The Impact of Tightening Tolerances in the U.S. Discrete Part Manufacturing Industry*, **NISTIR 4757**, Natl. Inst. Stand. Technol. (1992)

R. Tezza, A.C. Bornia, S.F. de Andrade, Measuring web usability using item response theory: principles, features and opportunities. Interact Comp. **23**, 67–175 (2011)

A. Viseu, Integration of social science into research is crucial. Nature **525**, 291 (2015). http://www.nature.com/news/integration-of-social-science-into-research-is-crucial-1.18355

T. Vosk, Measurement uncertainty: requirement for admission of forensic science evidence, in *Wiley Encyclopedia of Forensic Science*, ed. by A. Jamieson, A. A. Moenssens, (Wiley, Chichester, 2015). https://doi.org/10.1002/9780470061589.fsa1098

J. Weller et al., Development and testing of an abbreviated numeracy scale: a Rasch analysis approach. J. Behav. Decis. Making **26**(2), 198–212 (2013)

M. Wilson et al., A comparison of measurement concepts across physical science and social science domains: instrument design, calibration, and measurement. J. Phys. Conf. Series **588**, 012034 (2015)

J. Yang, W. Qiu, A measure of risk and a decision-making model based on expected utility and entropy. Eur. J. Oper. Res. **164**, 792–799 (2005)

Chapter 2
Measurement Method/System Development

Quality-assured measurement is a topic of burgeoning and increasingly multidisciplinary interest (Chap. 1) and demands increase continually for wider comparability and smaller uncertainties. Achieving this is particularly challenging when measurements are to be done over a wider scope, both in terms of scale (large and small, ranging from cosmological to nanoscales) as well as including more qualitative properties when tackling quality-assured measurement across the Social and Physical Sciences.

When embarking on a description of measurement in such challenging circumstances, it will be beneficial to exploit opportunities of observing that the measurement process is a specific example and subset of a production process, where the 'product' in this special case is the 'measurement result'. In the present chapter, obvious analogies will be drawn between designing production of entities—described in Chap. 1—and designing measurement systems for experiments.

In Fig. 2.1 is shown a quality assurance loop (for product, Sect. 1.3) in the special case where the 'product' is a measurement (Nilsson 1995). This measurement quality loop will provide the structure for basically the rest of the book. The present chapter—the first in this book to address predominantly measurement rather than product—will cover, so to say, the first quarter of the 'clock' of the loop; where the client issue of product demands is interpreted in terms of the corresponding measurement requirements. Modelling of a measurement system will be an important step (Sects. 2.2 and 2.4), as well as advice on designing experiments and measurement methods (Sect. 2.3) and finally validating and verifying them (Sect. 2.5). The chapter concludes with a couple of case studies before the reader is encouraged to continue their chosen example.

L. Pendrill, *Quality Assured Measurement*, Springer Series in Measurement Science and Technology, https://doi.org/10.1007/978-3-030-28695-8_2

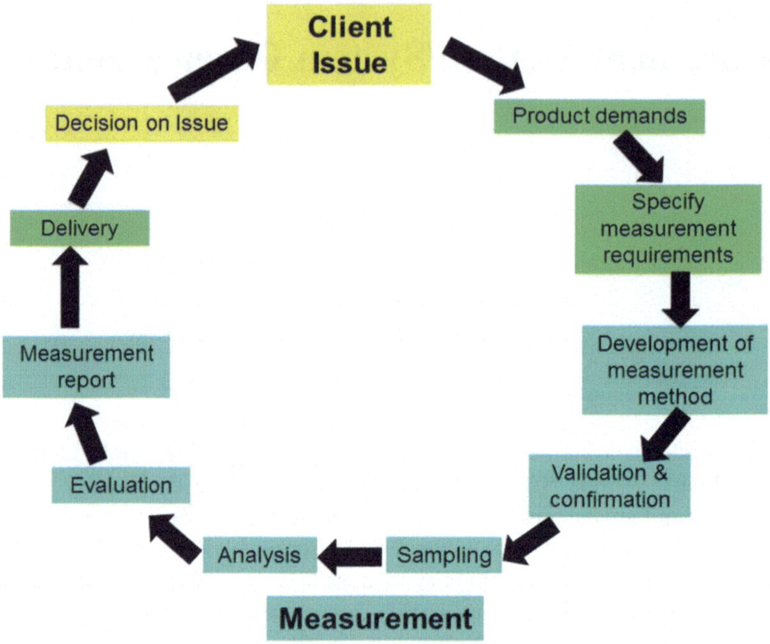

Fig. 2.1 Measurement quality loop (Nilsson 1995)

2.1 Design of Experiments

As with production processes (Chap. 1), measurement processes are never perfect and there will always be some dispersion in the observed entity value either for repeated measurements of one item or for measurements of a series of items. The overarching concept of 'accuracy' contains the respective concepts of precision and trueness, as defined in international documents such as ISO 5725 and the VIM and illustrated in Fig. 2.2, where the results of repeated measurements are shown as shots aimed at a bull's eye target.

2.1.1 Uncertainty and Risks. Fit-for-Purpose Measurement

As explained in Chap. 1, design of experiments in traditional statistics means the process of systematically varying controllable input factors to a 'manufacturing' process (in the broadest sense) so as to demonstrate the effects of these factors on the output of production. Analogously, much can be gained if one can plan measurements proactively for 'fitness for purpose', encompassing considerations—prior to

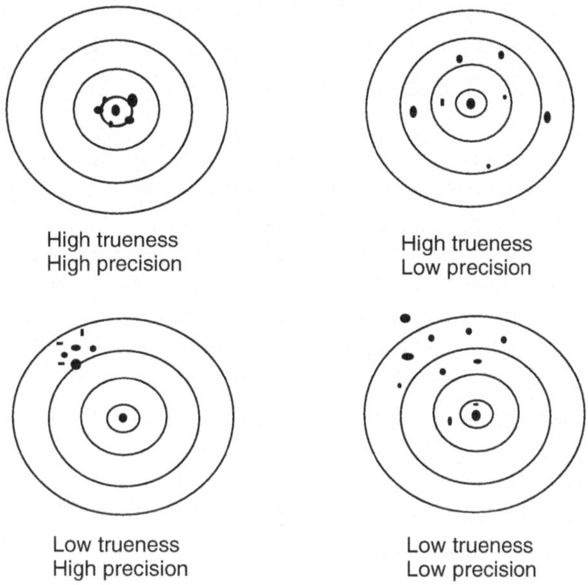

Fig. 2.2 Accuracy: precision and trueness (ISO 5725)

measurement—of aspects such as the measurement capability needed when about to test a product or process of a certain production capability.

The effects (cost and impact) of measurement at a certain level of uncertainty need to be compared with the consequence costs of incorrect decisions of conformity (JCGM 106:2012; Pendrill 2014c). The concept of 'uncertainty' here is a measure of the dispersion of unknown errors, reflecting a degree of indecision (see more about uncertainty in Chaps. 4, 5 and 6). Requirements for appropriately accounting for the consequences of measurement uncertainty when making decisions of conformity (somewhat analogously to the risk specifications added to ISO 9001—Sect. 1.4.8) have recently entered more prominently in the main written standards for conformity assessment, such as the latest version of ISO 17025:2017, which states:

> **7.7.1 Evaluation of Conformance** When statement of conformity to a specification or standard for test or calibration is requested, the laboratory shall:
>
> - document the decision rules employed taking into account the level of risk associated with the decision rule employed (false accept and false reject and statistical assumptions associated with the decision rule employed);
> - b) apply the decision rule.
>
> NOTE For further information see ISO/IEC Guide 98-4. (*JCGM 106:2012*)

Where ISO 9001 is a basic standard for *product* quality assurance, ISO 17025 can be regarded analogously as a basic standard for *measurement* quality assurance.

Measurement uncertainty in a test result—an apparent product dispersion arising from limited measurement quality—can be a concern in Conformity Assessment by inspection since, if not accounted for, uncertainty can (Pendrill 2014c):

- lead to incorrect estimates of the consequences of entity error,
- increase the risk of making incorrect decisions, such as failing a conforming entity or passing a non-conforming entity when the test result is close to a tolerance limit.

As with the analogous product risk handling (Sect. 1.4.5), measurement risk is not only a measure of the probability, p, of a risky measurement, but will also include a pragmatic measure of the impact, C, of an event when it occurs: that is, $risk = p \cdot C$ (Eq. (6.5)). Fit-for-purpose measurement means designing experiments to balance the costs of measurement against the risks of uncertain measurement (Pendrill 2014c). This will be considered further in this initial measurement chapter (Sect. 2.2.4) as well as when rounding off the whole book (Chap. 6) with final decisions about product.

2.1.2 Separating Production and Measurement Errors: Variability

One of the most difficult tasks in performing a measurement is to separate measurement dispersion from actual product variation. As illustrated in Fig. 2.3, the two types of scatter appear together in the displayed results and are easily confounded, even conceptually (Chap. 1).

At the same time, incorrect estimates of the relative magnitudes of product and measurement scatter can have serious consequences:

An underestimation of uncertainty in the measurement process (Fig. 2.3a) will imply a corresponding overestimation of product variation. That in turn can lead to an over-adjustment of manufacturing process and ultimately to larger variation in the final product, to the detriment of overall product quality.

On the other hand, an overestimation of uncertainty in the measurement process (Fig. 2.3b) will mean that actual product variation will go unnoticed, leading again to poorer product quality. One also risks unnecessary investment in new expensive measurement equipment or in re-training measurement engineers.

One way of reducing the risks associated with uncertainty is to set proactively limits on how large measurement uncertainty is allowed to be in relation to the object variability (Sect. 2.2). Reliability is one measure of this ratio—see further in Sect. 2.6 (eq. 2.13). Ultimately when appropriate (or fit-for-purpose (Thompson and Fearn 1996; Pendrill 2014c)) levels of uncertainty and associated risks of incorrect decisions are to be set when designing experiments, reference will need to be made to measures of impact for various stakeholder groups, as will be described in more detail in the latter steps of the quality loop (Sect. 1.3; Fig. 2.1) as covered in Chap. 6.

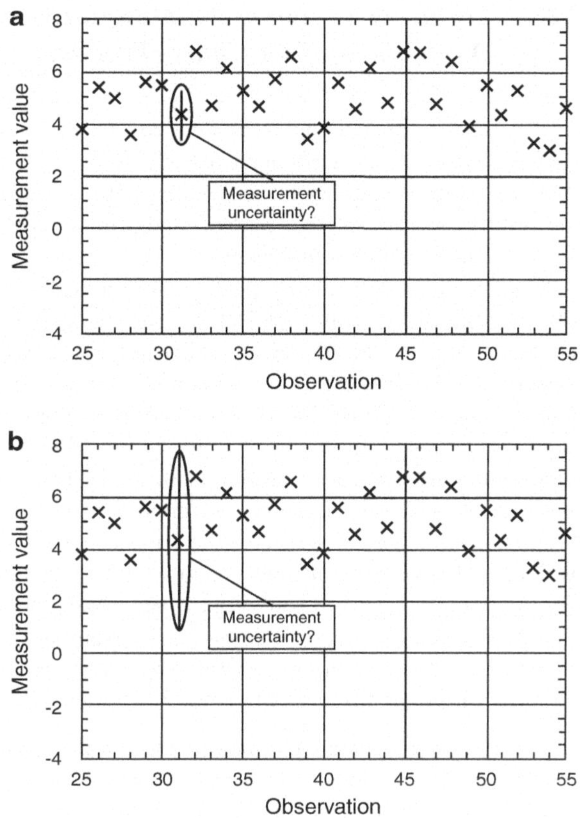

Fig. 2.3 (**a**) Underestimated uncertainty? (**b**) Overestimated uncertainty?

2.2 Specification of Demands on a Measurement System

Confidence in the measurements performed in the conformity assessment of any entity (product, process, system, person, body or phenomenon) can be considered sufficiently important that the measurements themselves will be subject to the steps (a) to (f) of the product quality loop (compare Sect. 1.3 with Fig. 2.1) as a kind of metrological conformity assessment 'embedded' in any product Conformity Assessment, as mentioned in Sect. 1.1. In order to assess an entity according to product specifications (Sect. 1.4.1), it will often be necessary to set corresponding measurement specifications. As with design of experiment, one can draw parallels between classical statistical process control of production (Montgomery 1996, Sect. 1.4.3) and metrological process control of measurement.

2.2.1 Different Variables Entering into Conformity Assessment of an Entity (Product)

A first step in formulating specifications is to define the constructs of interest. This discussion was introduced already in Sect. 1.4. As illustrated in Fig. 2.3, entity (product) and measurement scatter are often intertwined so that measurement conformity assessment (this chapter) and product conformity assessment (Chap. 1), so to say, meet at this initial stage.

From a philosophical perspective, the entity, which is the object of ultimate interest, can have both 'substance'—an unchanging essence of the entity in itself—as well as 'accidence'—that is, an appearance, where the two aspects of property can be differentiated conceptually but are inseparable (Kant, referenced in Kohlhase and Kohlhase 2008). In modern metrological terminology, the concept of 'definitional uncertainty' can be found in the VIM §2.27 definition: 'component of measurement uncertainty resulting from the finite amount of detail in the definition of a measurand'. In presenting an approach to measurement applicable to both the physical and social sciences, the current book will introduce (Fig. 1.1) a clear distinction between an attribute of the entity (e.g. task difficulty) and an attribute characterising the person (or 'probe') perceiving the entity (e.g. ability), where in our opinion this distinction is closer to the Kantian view than the VIM definition.

One can distinguish in general two types of entity-specific components of variation of the quality characteristic, Z,[1] of the object (Sect. 1.4.1):

- Variable Z_{specific}: Actual variation in the quality characteristic of one specific item of the product subject to conformity assessment (for example, changes arising when the item is used, for instance, handled in trade),
- Variable Z_{global}: Actual variation in the quality characteristic across a population of items of the product subject to conformity assessment (for example, each manufactured item will have a different value from the other items).

2.2.2 Metrological Characterisation and Specifications

Corresponding types of measurement-specific components of variation are:

- Variable Y_{specific}: Apparent variation in product, due to overall limited measurement quality when determining the value of the quality characteristic of one specific item of the product subject to conformity assessment,
- Variable Y_{global}: Apparent variation in product, due to overall limited measurement quality when determining the values of the quality characteristic of

[1]Where the distinction is important, a quantity is denoted with a capital letter, e.g. Z, while a lowercase letter, e.g. z, is used to denote the quantity value resulting from a measurement of that quantity.

a population of items of the product subject to conformity assessment (e.g. limited sampling).

Variations associated with limited measurement quality, expressed in terms of a measurement uncertainty PDF, $g_{test}(y)$ of the quantity $\xi = Y$ in the 'measurement space', i.e. the measurand, may partially mask observations of actual entity quality characteristic dispersion with PDF, $g_{entity}(z)$ introduced in Sect. 1.4.1, as illustrated in Fig. 2.3. To make clear the essential distinction between measurement variations and the quality characteristic variations that are the prime focus of conformity assessment, two different notations—*Y* and *Z*, respectively—have been deliberately chosen.

A construct specification equation was introduced in Sect. 1.4.3 (Eq. 1.2) when modelling the measurement object, the entity. Analogously, a design of experiment (DoE) approach in measurement similar to that in statistics can be performed where one would systematically vary controllable input factors (*x*, explanatory variables) to a measurement process and register the response, *y*. Allowance in this measurement DoE is made for both (1) variability (dispersion)—dealt with by performing analyses of variance, risk assessment and optimised uncertainty methods—and to (2) bias (location)—dealt with by performing metrological calibration.

The two components of measurement accuracy, depicted as precision and trueness in Fig. 2.2, can be determined both under conditions of *repeatability (r),* defined as repeated measurements with *one and the same* measurement system during a relatively short time interval, and of *reproducibility (R)*, in which each new measurement in principle is made with a *different* measurement system, i.e. ideally a new measurement object, new instrument, new operator and so on (ISO 5725). Further description of a measurement system can be found later in this chapter while Chap. 4 (Fig. 4.3) describes further analysis of measurement scatter under different conditions.

For some tasks where some local effect on the measured object is of interest, repeatability conditions may suffice. But although precision in those cases will appear higher (less scatter), if one is interested in investigating a more universal understanding of measurement quality, it will be worthwhile investing more resources in a full-scale reproducibility study which will have more scatter but hopefully will cover most eventualities. For instance, it is usually the case that a manufacturer wants to be sure that all his products, whenever and wherever produced, and not just one, shall have assured quality—in which case measurements to test product should be quality-assured under conditions of reproducibility. Examples will be given in Chap. 4.

When considering trueness, an immediate means of checking that a measurement instrument is reading correctly can be done by the measurer himself, measuring a local check standard at regular intervals with the instrument (ISO 10012). The metrological process of calibration is more generally fundamental to ensuring freedom from bias in measurements by making any measurement result traceable, through an unbroken chain of calibrations in a hierarchy of comparisons, to a primary reference standard (etalon[2]) for each measurement quantity. As described in detail in Chap. 3, calibration enables any measurement to be quantitatively comparable to other measurements made on other occasions by different people and is a prerequisite for ensuring that the tested product will have corresponding properties of interoperability. In the social sciences the relevant concept is intersubjectivity (Gillespie and Cornish 2010) and the concepts illustrated in Fig. 2.2 are sometimes presented where precision and trueness are replaced, respectively, by reliability and validity (Wenemark 2017). Examples of standards referring to calibration include: 'Use of certified reference materials' ISO Guide 33 and 'Linear calibration using reference materials' ISO 11095.

Metrological confirmation, in the terminology of ISO 10012 (Sect. 4.2), covers methods of checking that critical functions and performances when producing measurement results satisfy specifications and is in a sense analogous to ISO9000 quality assurance for more general products. Measurement equipment shall have those metrological characteristics demanded of the application at hand (for example concerning accuracy, stability, measurement range and resolution). This means that one starts a measurement task by evaluating what is to be measured, and thereafter choose suitable equipment. A metrological characteristic is any distinguishing feature (of a measurement system) which can influence the results of measurement (ISO 10012:2002 §3.4).

One established area where conformity assessment of measurement instruments is regularly performed is legal metrology. In that field, instrument testing is used widely as a proxy for the control of actual measurements of goods and utilities in society, as covered by the EU Directive Measuring Instrument Directive (MID)—both by type approval, initial and subsequent verification (Källgren and Pendrill 2006; EU commission 2014). Conformity assessment procedures in legal metrology can be regarded as a prototype for more general conformity assessment (EU commission 2018). Typically, alongside more qualitative attribute requirements, such as inspection of correct instrument labelling and unbroken sealing of instruments, measurement specifications are also set by variable in terms of maximum permissible errors (*MPE*), both for the main characteristic (e.g. indication of an electricity energy meter) as well as of any influence quantity (e.g. level of disturbing electromagnetic field, in EMC testing) to be tested through quantitative measurement.

[2]'etalon': a French word usefully distinguishing a measurement standard from a written standard (norm).

2.2.3 *Separating Object and Instrument Attribute Estimation. Measurement System*

To deal with the challenging task of separating unintentional dispersion due to limited measurement quality from the sought-after object variability, an important tool is conformity assessment of *the other main elements* of the measurement system (instrument, environment, operator, method), depicted in Fig. 2.4. This is a logical next step (Sect. 1.3 step b) and Fig. 2.1), having dealt with the conformity assessment of the measurement object—i.e., the first (and last) focal point in the quality assurance loop opened in Chap. 1. And bear in mind that in measurements in the social sciences, the instrument of the measurement system may well be a human being (Fig. 1.1b).

An initial step—perhaps one of the most important in evaluating measurements—is to formulate a valid model of the measurement system intended for use, say, in conformity assessment of a product, where one attempts to draw a version of Fig. 2.4 for the actual set up; including as appropriate one or more instruments; operators; etc. Making such a first 'sketch' of the measurement scenario is analogous to a detective making a quick summary of a 'crime scene'; identifying the most salient features and 'prime suspects', to be used in later investigations when searching for sources of error and uncertainty. If one unintentionally misses a key element of the measurement system at hand, no amount of advanced statistics will compensate. (Validation of measurement methods is dealt with in Sect. 2.5). It will also become obvious that measurement system analysis (MSA) is not only useful in traditional engineering and scientific metrology, but also plays, surprisingly, an essential role in describing measurement in the social sciences (Sect. 1.2). MSA faithfully describes the observation process, that is, how measurement information is transmitted from the measurement object, via an instrument, to an observer and as such is a particular example of a system for communication as dealt with in information theory (Chap. 3).

Measurement system analysis (MSA) is an essential tool in configuring the necessary measurement resources to have properties commensurate with the task ahead (Bentley 2005; AIAG 2002; ASTM 2012; Loftus and Giudice 2014; Rossi 2014). MSA has its roots in the statistical design of experiments, as described

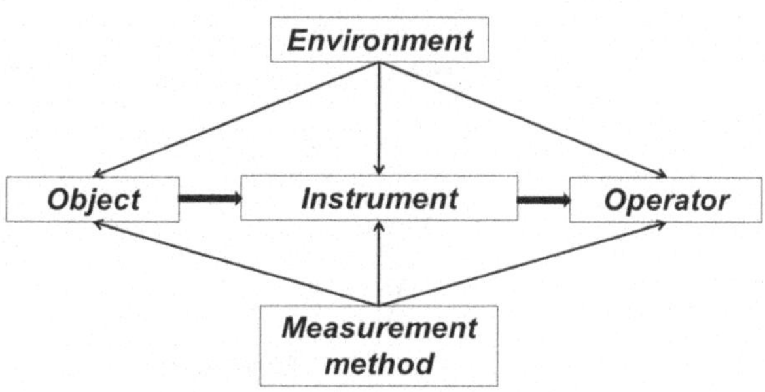

Fig. 2.4 Measurement system

by Montgomery (1996) where system output, Z, is some function of in general a number of controllable inputs, X, in the presence of several uncontrollable inputs, ε. Industrial design of experiments, where production is designed, is applied in the present case to designing measurements for 'production' of measurement results.

Measurement system analysis (MSA) enables the separation of instrument and measurement object attributes values necessary for any measurement system, such as when determining the mass of a weight in terms of the calibrated response of a weighing instrument. As explained in Chap. 1, we single out measurement as the specific case of a construct specification Eq. (1.2) where it is the 'response' of a measurement system which is of interest.

The process of restitution (Rossi 2014, Sect. 5.2) yields an estimate of the measurand, $z = S$, the required quantity of the measurement object (e.g. a mass) from the instrument response, $y = R$, (e.g. of a weighing scales) using the measurement construct specification equation. Note that the four variables z, S, R and y, introduced here, are in general distinct and have specific values in the restitution process. Restitution in the simplest case is done with the expression:

$$z = S = K_{\text{cal}}^{-1} \cdot R = K_{\text{cal}}^{-1} \cdot y \tag{2.1}$$

where the instrument sensitivity $K = K_{cal}$ is already known as a result of calibration—i.e. from the measured response of the instrument to a known stimulus, as described in more detail in Sect. 2.4. Without that separation, dispersion in the sought object attribute would be masked by instrument dispersion, thus limiting measurement quality in terms of both measurement precision and trueness.

One can obtain estimates for a typical set of metrics or indicators for a measurement system exemplified in Fig. 2.5, from a series of experiments with a measurement system used under repeatability and reproducibility conditions (Sect. 2.2.2), based on an analysis of variance approach (ANOVA). With reference to the concepts

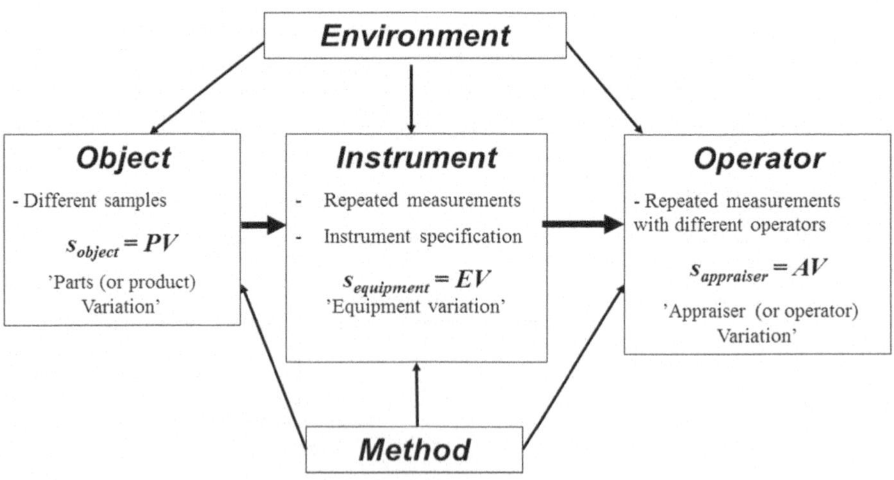

Fig. 2.5 Analysis of variance of a measurement system (AIAG 2002; ASTM 2012)

illustrated in Fig. 2.1, accuracy can be specified in terms of precision and trueness under both repeatability and reproducibility conditions (ISO 5725) for each MSA element.

Note that an important contribution to overall variance of the measurement system can be expected from the measurement object itself, as captured with the term *PV* (= parts or product variation). This provides the link between our immediate measurement concerns of the present chapter and the primary entity (product) studies introduced in Chap. 1 and illustrated in Fig. 2.3. According to the so-called Gauge *r & R* approach used frequently in for example the automobile industry, contributions to the overall measurement variance include also *EV*—i.e. equipment (or instrument) variance as well as *AV*—i.e. appraiser (or operator) variance.

Details about how to calculate the various variation measures shown in Fig. 2.5 can be found for instance in the appendices of ASTM (2012). Typically, a table such as given below (Table 2.1) is filled in for n parts (objects), m repeats per object and p operators. The corresponding measurement model is that the response output, y, of the measurement system is expressed as:

$$y_{jik} = z_j + o_i + w_{ji} + \varepsilon_{jik};\quad j = 1\ldots n, i = 1\ldots p, k = 1\ldots m \tag{2.2}$$

where the attribute (stimulus input) value, z, of the measurement object can be distributed about a mean value $\bar{z} = \mu$ for the series of parts (objects, j) manufactured to nominally the same value. The other variables on the right-hand side of Eq. (2.2) denote variations distributed about a mean value of 0 (zero) for each element of the measurement system (o for operator, i, variation; w_{ji} denotes part-operator interactions), and where ε_{jik} represents repeatability variations.

Calculation of the various variation measures follow the usual Analysis of Variance (ANOVA) formulae which may be found in any standard statistics book, such as Montgomery (1996). To the extent that these various variation terms are not compensated for, the residual dispersion for each factor will constitute a component

Table 2.1 Template for recording measurement system responses, y, and for ANOVA

Trials	Part 1	Part 2	Part 3	Part 4	...
Operator A1	y_{111}	y_{211}	...	–	–
Operator A2	y_{112}	...	–	–	–
Operator A3	...	–	–	–	–
Average A		–	–	–	–
Range A		–	–	–	–
Operator B1	y_{121}	–	–	–	–
Operator B2	–	–	–	–	–
Operator B3	–	–	–	–	–
Average B	–	–	–	–	–
Range B	–	–	–	–	–
...	–	–	–	–	–

of measurement uncertainty, u, in the overall response of the measurement system (JCGM 100:2008, GUM).

Variations from other elements (e.g. instrument or environment) can be added to the measurement model at will as extra factors on the right-hand side of Eq. (2.2). It is also implicitly assumed in formulating Eq. (2.2) that the sensitivity, K, of the measurement system (appearing in Eq. (2.1)) is constant and equal to 1 (one); that is, the response y is a simple linear function of the stimulus z. (In the next sections we will generalise the measurement system model to account for various effects, such as non-linearity, interferences, etc. as well as allow the response to be ordinal, as is of interest in measurements in the social sciences, for instance).

2.2.4 Measurement System Specifications

Once again drawing parallels with product conformity assessment (Sect. 1.4.2), measurement specifications can be set prescriptively on characteristics which appear on either side of the construct specification equation (Eq. (1.2)) describing the outcomes.

Metrological specifications for the properties of the measurement system—either for the system as a whole or each element (object, instrument,...) of the system—are of two distinct kinds:

- limits on maximum permissible measurement uncertainty (or, equivalently, minimum measurement capability) when testing product,
- limits on maximum permissible error in the indication of the measurement equipment/system intended to be used in the measurements when testing product.

MPE Instrument

Setting limits on maximum permissible error (*MPE*) in the indication of the measurement equipment/system intended to be used in the measurements when testing product can be viewed as a special case of general conformity assessment (where the latter was described in Chap. 1).

The entity subject to metrological conformity assessment is the measurement equipment/system and the quality characteristic can be the:

- indication of the display of the measurement instrument,
- error associated with the chosen measurement method, operator, etc.

Setting an *MPE* on the measurement system is one way of ensuring that, when measurements are actually performed when testing product, requirements on maximum permissible measurement uncertainty (*MPU*) are likely to be satisfied and commensurate decision risks are reduced to an acceptable level: Whether they will or not depends not only on instrument specifications but also on the actual metrological performance when measuring.

A main focus when setting specifications on a measurement system in this context will be to ensure that, as far as possible, apparent dispersion due to limited measurement capability will be small, so as not to mask actual product errors, which of course are the end target.

Capability Factors

Specifications on measurement process capabilities can be made analogously to specifying production process capabilities.

Limits on both *process* and *measurement capabilities* are traditionally set in terms of certain factors—C_p and C_m, respectively—which are fractions of the same product tolerances (specification limits, U_{SL} and L_{SL}, for the magnitude of a characteristic of an entity) and maximum permissible product value errors, e.g. $U_{SL} - L_{SL}$.

Analogous to process capability (Sect. 1.4.2), a measurement capability index, C_m, can be defined in terms of estimated measurement variations as: $C_m = \frac{U_{SL} - L_{SL}}{M \cdot u_m}$ with standard measurement uncertainty u_m and typically $M = 4$ (corresponding to a coverage factor, $k = 2$ and 95% confidence).

In the context of measurement system analysis illustrated in Fig. 2.5, key indicators of measurement system performance are (ASTM 2012):

- 'Repeatability & Reproducibility'

$$GRR = \sqrt{EV^2 + AV^2} \tag{2.3}$$

- 'Total variation' $TV = \sqrt{GRR^2 + PV^2}$

These expressions (Eq. (2.3)) define *GRR* and *TV* in terms of the factors *PV, EV* and *AV* defined in Fig. 2.5.

Limits on Capability Factors

Typical requirements on variances intended to minimise the risk that limited measurement quality may mask assessment of intrinsic product dispersion can be set in various ways.

The *maximum permissible uncertainty* or 'target uncertainty', $MPU = 1/C_{m,\min}$ in terms of a corresponding *minimum measurement capability*, $C_{m,\min}$. In various sectors of conformity assessment, different limits on measurement capability have become established, with $C_{m,\min}$ ranging from typically 3 to 10.

Table 2.2 Acceptance criteria for measurement systems

$r \,\&\, R < 10\%TV$	Acceptable
$10\% < r \,\&\, R < 30\%$	Barely acceptable
$r \,\&\, R > 30\%TV$	Measurement system needs to be improved.

A common limit to ensure that measurement quality variations are small is $\frac{u_m}{s_p} < 30\%$, as in Measurement System Analysis (MSA) in the automobile industry, for instance (AIAG 2002; ASTM 2012) (Table 2.2).

The ISO 10012 standard refers to a metrological confirmation system whose purpose is to limit the risk of unacceptable errors in measurement equipment. Errors associated with calibration should be as small as possible: 'For most measurement tasks calibration errors should be less than a third and preferably only one tenth of the error permitted when using the confirmed equipment' (Chap. 4).

In psychometrics a measurement reliability coefficient (calculated with Eq. (2.11)) of 0.8—corresponding to a measurement uncertainty of about one-half of the measured dispersion—is considered acceptable for so-called high stakes testing (Linacre 2002). A factor one-half is also a pragmatic limit (Pendrill 2005, Sect. 3.4.3) as discussed further in Sect. 4.3.2.

A general question, as yet apparently unanswered in the literature, is the following: in the context of conformity assessment of an instrument (or measurement system), we have seen in this section how a maximum permissible measurement error, MPE_{measure}, can be specified. In the context of product conformity assessment, a maximum permissible product error, MPE_{entity}, is specified as discussed in Chap. 1. The question is: what is the relation between these two *MPE*s—product and measurement permissible errors?

Many of the abovementioned rules have a certain element of arbitrariness and limits vary, often with little motivation as to the actual consequences of incorrect decision-making in conformity assessment. Questions of appropriate rules for decision-making in conformity assessment with due account of measurement uncertainty raise questions about fit-for-purpose measurement (Sect. 2.1) which ultimately can be resolved by economic considerations (Sect. 6.4).

2.2.5 Examples of Measurement System Demands, Both Quantitative and Qualitative. Pre-packaged Goods

To illustrate how requirements can be specified at the present stage of design of experiment concerning the performance of the measurement systems to be used in testing, the following Table 2.3 gives an example for the case of pre-packaged goods. The reader is encouraged to perform a similar exercise by filling in the templates provided at the end of this chapter for their product of choice.

Table 2.3 Demands on measurement system and methods, based on product demands §E1.3

	Your answers... Coffee powder pre-packaged
For as many as possible of each product test (A, B, C, D) under §E1.3, specify demands on measurement system and methods:	B. function. The packet should contain as promised: - correct amount powder - 500 g
- Minimum production capability ($C_{p,\mathrm{min}}$)?	0.5 ($s_p = 10$g, $MPE = 15$ g) 99% confidence
- Minimum measurement capability ($C_{m,\mathrm{min}}$)?	2.5 ($u_m = 3$g, $MPE = 15$ g) 95% confidence
- Maximum permissible measurement uncertainty (*MPU*)?	3 g
- Maximum permissible measurement (*MPE of instrument/system*)?	0.2 g
Other demands:	B. function. One experiences a certain 'prestige' when serving this mark of coffee: - Measure of perceived 'prestige': 80%, at least 60% (customer survey)
- Minimum production capability ($C_{p,\mathrm{min}}$)?	0.5 ($s_p = 10\%$[1], $MPE = 20\%$) 99% confidence
- Minimum measurement capability ($C_{m,\mathrm{min}}$)?	1 ($u_m = 5\%$, $MPE = 20\%$) 95% confidence
- Maximum permissible measurement uncertainty (*MPU*)?	5%
- Maximum permissible measurement (*MPE of instrument/system*)?	2% (may be difficult!)

In the case of pre-packaged goods, requirements range from quantities—such as the mass of goods in each packet, which are quantitative with a focus on the technical capabilities of the production processes, for instance the performance of filling machines—to properties such as the perception of the product, where more qualitative properties such as 'prestige' are of interest (Sects. 1.5 and 4.5.2).

At the present design-of-experiment stage, a suitable measurement system (as depicted in Fig. 2.4) will have to be configured for each quality characteristic of interest. Often quantitative information will be associated with (A) Test of non-functional characteristics of product; is the product 'correctly' made? while qualitative information (B) Test of product function; is the product 'right'? (Sect. 1.4.4).

Section 1.5.3 introduces the four different kinds (A, B, C, D) of entity (product) tests in general, as well as giving examples of these for the chosen example of pre-packaged goods presented here. For the straightforward task of determining adequate filling by mass of the goods packets, one would obviously plan to procure a good quality weighing machine. For the more qualitative, but no less essential, determination of the perceptual goods properties to satisfy the consumer, one could envisage a measurement system where Man acts as the measurement instrument (Sect. 1.2, Berglund et al. 2011; Pendrill 2014a, b). Measurements are then made on each side of the system triangle shown in Fig. 1.5. A process of restitution (Eq. (2.1)) will be required to yield separate measures of the item and person attribute values listed in Table 1.1. See Sect. 2.4.4 for further discussion. Actual measurements will be exemplified in Chap. 4.

The same reservations about handling the more qualitative information about product (mentioned in Sect. 1.2) also apply in the present case of specifying demands on measurement systems and methods, as exemplified in Table 2.3. Is there any sense, for instance, in specifying limits (Fig. 1.3) on the location or dispersion of perceived 'prestige' which usually lies on an ordinal scale, where distances between different marks on the scale (Fig. 1.1) are not known exactly?

An introduction to handling more qualitative information will be given in the case study—Sect. 2.7.2—presented at the end of this chapter, which will be followed up with general procedures described in detail in later chapters of this book.

2.3 Choice and Development of a Measurement Method

The choice of measurement method—as one element of the measurement system shown in Fig. 2.4—is an essential next step, having set requirements on the required metrological performance of the measurements to test product according to specification. The development of a new method is based on an error budget modelled in a similar way to that used above (Eq. (2.2)) when describing other elements of the measurement system.

The final measurement method will include specification of the key factors which can affect the measurement result. A measurement method can be described in terms of

- measurement principle,
- procedures,
- equipment,
- target values and tolerances (specifications) for operational performance and functionality.

Later on (Sect. 2.5), a description of evaluation and validation of a measurement method will be given in detail. Alongside so-called 'accuracy' experiments (ISO 5725), and as a complement to experimental testing, modern computing can enable simulation to be used to advantage for example in this method development work.

2.3.1 *Definition of Test Problem, Based on Test Requirements*

In elucidating the measurement principle, one can draw parallels between the four different kinds of entity tests (Sect. 1.5.3) and the corresponding test of measurement systems intended for use when assessing conformity in the metrological context, as can be listed simply by replacing the word 'product' with 'measurement system' in the following list (identified in Chap. 1):

A Test of non-functional characteristics of measurement system; is the system 'correctly' made?
B Test of measurement system function; is the system 'right'?
C Initial verification,
D Subsequent verification.

To provide a complete picture of the tests to be performed, it will in general be necessary to consider both functional and non-functional testing of measurement systems for each separate quality characteristic—both functional and non-functional—of the product subject to conformity assessment. In the next section, measurement system analysis will provide the tools necessary for this purpose. The chapter concludes with examples. The reader is encouraged to complete the exercises provided at the end of this chapter (Exercise E2. Definition of Measurement Problem).

2.4 Measurement System Analysis (MSA)

Essential insight can be gained by extending traditional metrological concepts through developing an operational model of a measurement system (Sect. 2.2). Others have already pointed out that this perspective has not been emphasised enough, even in traditional metrology let alone quality-assured measurement in the social sciences: 'One major difficulty for an extensive application of GUM's principles… [has been an].. almost total absence of (the notion of a measurement system) in The Guide to the evaluation of Uncertainty in Measurement (GUM) [JCGM 100]', according to Rossi and Crenna (2016). In our work, it has turned out that a measurement system analysis approach is also essential in describing measurements in the social sciences (Sect. 1.1, Pendrill 2018).

2.4.1 *Measurement Model*

In the formulations of measurement system models, such as Eq. (2.2) in the context of MSA as well as method accuracy experiments (Sect. 2.5), assumptions are made that for instance the sensitivity of the measurement system (*K*, appearing in

Eq. (2.1)) is constant and equal to 1 (one); i.e. the response y is a simple linear function of the stimulus, z. If such simplifications are valid, it is straightforward to deduce the sought-after value of the quality characteristic of the measurement object by simply taking the displayed response, y, of the measurement system as equal to z, in accord with Eq. (2.2).

In general, of course, simplifications of this kind are not the whole story and one has to be prepared to describe the measurement system with models which can account for a number of effects, such as non-linearity, interferences, etc. as well as allow proper treatment of the response if it happens to be ordinal, as is of interest in measurements in the social sciences, for instance.

2.4.2 *Static Functional Characteristics of Measurement Systems*

As part of the process of elaborating the initial simple restitution model Eq. (2.1), the most important measurement system functional characteristics can be listed in Table 2.4 in terms of the response, R, of the system $= O$, an output signal—for a given input, I, i.e. stimulus S (based on Bentley 2005, Chap. 2).

In general, these characteristics are evaluated by investigating the output response of a measurement system over a range of known input stimulus levels. It is useful to make plots such as exemplified in Fig. 2.6, of measurement system response output, O, versus input, I, for modifying and interfering (with bias b) environmental 'nuisance' effects.

The various characteristics (Table 2.4) of the measurement system can be summarised with the model shown in Fig. 2.7, and mathematically with the expression (Bentley 2005):

Table 2.4 Static characteristics of measurement

	Input	Output
Range	I_{MIN} to I_{MAX}	O_{MIN} to O_{MAX}
Swing	$I_{MIN} - I_{MAX}$	$O_{MIN} - O_{MAX}$
Non-linearity	I	$O = f(I)$
Sensitivity	I	$K = \frac{dO}{dI}$
Resolution	dI	$dI_{res} = dI_{max}$ when $dO = 0$
Wear	I	$O(t_2) \neq O(t_1)$
Error limits	I	$O - h < O_0 < O + h$
Environmental Effects		$O' = f'(I)$
(a) Modifying	I	$K' = \frac{dO'}{dI} \neq \frac{dO}{dI}$
(b) Interfering		$O' = O + b$
Hysteresis	I_{up}, I_{down}	$O(I_{up}) - O(I_{down})$

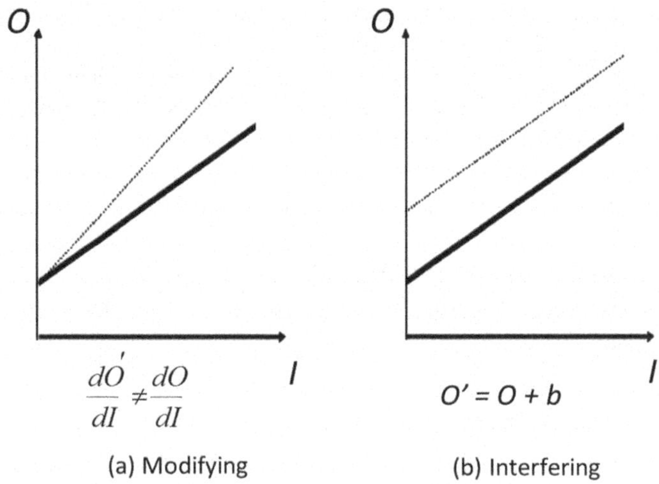

Fig. 2.6 Measurement system response plots of output, *O*, versus input, *I*, for (**a**) modifying and (**b**) interfering (with bias *b*) environmental effects

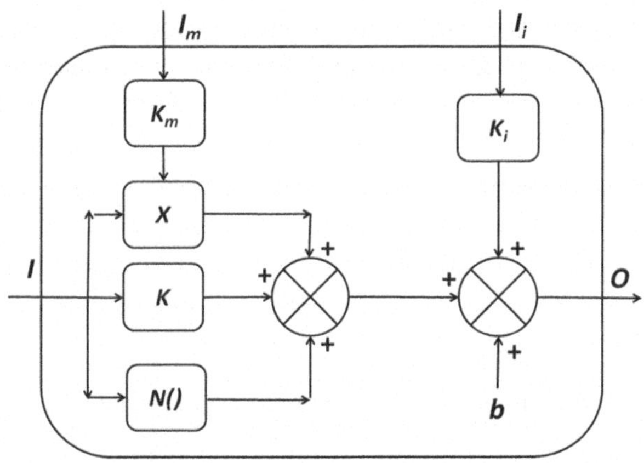

Fig. 2.7 Model of measurement system (The block diagram symbols in Fig. 2.7 are explained in (Bentley 2005, p. 7, Fig. 1.4). For instance, 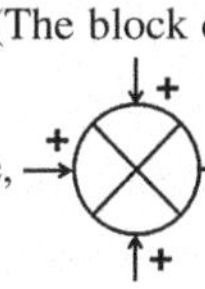 indicates an output to the right which is the sum of the three inputs)

$$O = K \cdot I + N(I) + K_M \cdot I_M \cdot I + K_i \cdot I_i + b \tag{2.4}$$

Sensitivity = K; Non-linearity = $N(I)$; bias = b; Modifying disturbance = I_M, with sensitivity = K_M; Interfering disturbance = I_i, with sensitivity = K_i.

Models of measurement systems such as Eq. (2.4) and Fig. 2.7 have of course to be tested, validated and verified, as will be discussed in Sects. 2.5 and 2.6. The same basic model is expected to be applicable to both traditional physical and engineering measurement systems as well as cases, for instance in the social sciences, where Man acts as a measurement instrument.

2.4.3 Measurement System Modelling As a Chain of Elements: Signal Propagation in a Measurement System

In the words of Guilford (1936):

> … all measurements are indirect in one sense or another. Not even simple physical measurements are direct, as the philosophically naïve individual is likely to maintain. The physical weight of an object is customarily determined by watching a pointer on a scale. No one could truthfully say that he "saw" the weight…

Bentley (2005, Fig. 1.3) presented models of a measurement system consisting of a chain of elements, in general consisting of a mixture of up to four basic kinds: (a) sensing; (b) signal conditioning; (c) signal processing; and (d) data presentation. An illustrative example given by Bentley is of a temperature measurement system where the sought-after object temperature is T_1. Each element is typically described as follows.

(a) *sensing*: thermocouple with reference temperature T_2 and producing an output signal

$$E_{T_1 T_1} = b_0 + a_1 \cdot (T_1 - T_2) + a_2 \cdot \left(T_1{}^2 - T_2{}^2\right)$$

(b) *signal conditioning*: amplifier producing an output signal

$$V = K_1 \cdot E_{T_1 T_1} + b_1$$

(c) *signal processing*: analogue-to-digital converter producing an output in digital bits

$$n = K_2 \cdot V + b_2 \tag{2.5}$$

(d) *data presentation*: microcomputer with display, showing output

$$T_m = K_3 \cdot n + b_3$$

As can be seen, the output of each element provides the input to the next element in the chain.

In this book an important fifth category of measurement system element will be added, namely (e) a decision-making element. Most measurements are not made solely for the sake of measurement, but because decisions are to be made about something (the entity, Chap. 1) based on the measurements:

(e) *decision-making*: algorithm producing an output on a categorical scale: the result of a decision, such as the binary, dichotomous response to e.g. the question 'is the temperature T_m below or above tolerance T_{SL}?'

$$R = \begin{pmatrix} 0 \\ 1 \end{pmatrix} \textit{if} \begin{bmatrix} T_m \leq T_{SL} \\ T_m > T_{SL} \end{bmatrix} \tag{2.6}$$

or a polytomous response distributed over a number of categories. Typically, decisions can be of two kinds, as in psychophysics (Iverson and Luce 1998):

- *identification*: T_{SL}in (Eq. (2.6)) is a specification limit for the quality characteristic of the entity being assessed for conformity,
- *choice*: $T_{SL} = T_{m'}$ in (Eq. (2.6)) where $T_{m'}$ is a second (e.g. prior) measurement result.

2.4.4 Performance Metrics of Measurement Systems

Traditional metrics with which the performance of a measurement system is rated are typically expressed in terms of measurement error: how close the system response is to the 'correct' value?

Propagation of measurement bias and dispersion can be modelled with the following two expressions, respectively:

- *Accuracy (trueness) = measured value − true value = system output − system input,* $O_j - O_{j-1}$ such as $T_m - T_1$ in the thermometer example
- *Accuracy (precision)*:

$$\sigma_{O_j}^2 = \sigma_{I_{j+1}}^2 = \left(\frac{\partial O_j}{\partial I_j}\right)^2 \cdot \sigma_{I_j}^2 + \left(\frac{\partial O_j}{\partial I_{M_j}}\right)^2 \cdot \sigma_{I_{M_j}}^2 + \ldots \tag{2.7}$$

including as many terms for each element *j* of the measurement system as appear in Eq. (2.4), while assuming no correlation between the different elements. Each element of the measurement system will include a variety of measurement quantities.

While many measurement systems deliver responses on quantitative, continuous scales, in some cases, such as the analogue-to-digital converter (Eq. (2.5)) and the decision-making algorithm (Eq. (2.6)), the outputs will often be on discrete scales. For categorical response cases (Fig. 1.2), including the important decision-making response (Eq. (2.6)), it is not immediately obvious whether expressions such as of accuracy (Eq. (2.7)) can be applied at all, since the exact mathematical distances between different categories cannot be assumed to be known. For these categorical responses, measurement system 'accuracy' will be identified with decision-making ability:

- Accuracy (decision-making) = response categorisation – input (true) categorisation (2.8)

where P_{success} is a metric of measurement system performance in terms of the probability of making the 'correct' decision.

For a simple binary decision (Eq. (2.6)), a correct decision is described as assigning the response to the category at the output of the measurement system corresponding to the 'correct' category of the measurement entity at the input to the measurement system. Analogous to the usual measurement error (Eq. (2.7)), the closer the categorisation, the greater the 'accuracy', measured in terms of P_{success} (Eq. 2.8).

Bashkansky et al. (2012) describe an 'accurate' system as one in which the off-diagonal elements (α and β, the risks of type-1, respectively, type-2 decision errors) of the (binary) confusion matrix $\begin{pmatrix} 1-\alpha & \alpha \\ \beta & 1-\beta \end{pmatrix}$ are minimised (where $P_{\text{success}} = 1 - \alpha$ or $1 - \beta$). Akkerhuis et al. (2017) take a similar approach, where measurement error is expressed as a misclassification, statistically evaluated in terms of the misclassification probabilities, α and β (or their complements sensitivity and specificity) for binary tests ranging broadly from visual quality inspections of industrially manufactured parts to diagnostic and screening tests in medicine. In the example 1 of VIN §3.9 *Examination uncertainty* given by Nordin et al. (2018): 'The reference nominal property value is "B". The nominal property value set… of all possible nominal property values is {A, B}. For one of 10 examinations … the examined value differs from 'B'. The examination uncertainty is therefore 0.1 (10 %)'. Again, misclassification probabilities, α and β, are considered as accuracy measures.

But performance metrics such as P_{success}, α or β often belong to the 'counted fraction' kind of data and in general are of ordinal, rather than fully quantitative nature, not directly amenable to regular statistics (Sect. 3.5.1). There is also the same task of separating the instrument factor from the sought-after object factor even in qualitative, categorical responses of the measurement systems. How to deal with this will be considered further in the next section, Sect. 2.5.3, a pre-packaged example in Sect. 2.7.2 and later in the book: in Sect. 3.5 and in depth in Chap. 5.

2.4.5 Restitution

It will be possible to predict a response of the measurement system to any arbitrary input stimulus value once the various coefficients in Eq. (2.4) have been evaluated by experiment in a calibration and test procedure made over a range of known input values (Rossi 2014).

If the input signal is measurement information on a quantitative interval or ratio scale from the measurement object, then the sought-after value of the quality characteristic of the object can be deduced by a restitution process in which Eq. (2.4) is inverted to estimate $I = S$ in terms of the other terms; assuming of course that system factors, such as the sensitivity K, remain unchanged since calibration was performed. A simple example is a measurement system where the instrument sensitivity $K \neq 1$ and there is an offset (bias), b, in the output, O. The formula for restitution of an unknown input, I, that is the stimulus value, S, of the measurement object in this case is:

$$z_j = S_j = \left(\frac{R-b}{K}\right)_j = \frac{y_j - b_j}{K_j} \tag{2.9}$$

which might need to be evaluated individually at every input level if either sensitivity and/or bias vary with level.

For the categorical responses of for instance the decision-making elements (Eq. (2.8)), restitution takes an analogous form: A correspondingly simple example could be a measurement system where instrument sensitivity is some measure of how much a human being responds to a certain stimulus and where there can be a degree of bias when making decisions expressed as a decision-making accuracy—in terms of P_{success}, that is the probability of making a correct categorisation, as in Eq. (2.8). The Rasch (1961) measurement model, mentioned in Sect. 1.2.3 and $z = \theta - \delta = \log\left(\frac{P_{\text{success}}}{1-P_{\text{success}}}\right)$ Eq. (1.1), can be applied in the first approximation to transform the ordinal, 'counted fraction' data (Sect. 3.5.1), P_{success}, onto the more quantitative scale for θ and δ. But in general one needs to take account of the fact that arguably a human being acting as a measurement instrument is often quite less reliable than say a traditional instrument in engineering or physics. For more generality therefore, the basic Rasch model (Eq. (1.1)) may need extending to include both guessing, b, and varying instrument discrimination, ρ. The stimulus constituting the input to the measurement system can for instance be modified during transmission through the instrument to $\rho \cdot (\theta - \delta)$, where the discrimination ρ is taken to be some function of the instrument sensitivity. This expression can be handled according to the procedure suggested by Humphry (2011) for handling measurement units in the logistic measurement function (Eq. (3.7)).

From the 3PL logistic model (Partchev 2004):

$$P_{\text{success}} = b + (1 - b) \cdot \frac{e^{\rho \cdot (\theta - \delta)}}{1 + e^{\rho \cdot (\theta - \delta)}} \tag{2.10}$$

an expression for the restituted stimulus for a qualitative measurement system can be derived as:

$$z = S = \theta - \delta = \log \left[\frac{\frac{P_{\text{success}}}{1-b} - b}{1 - \frac{P_{\text{success}}}{1-b} - b} \right] - \log(\rho) \tag{2.11}$$

A complete expression can be derived by modelling relations between the manifest response of the measurement system and a stimulus input from the measurement object. A structured equation model (construct specification Eq. (1.2)) can be formulated, where multivariate analyses are to be made in cases where there is significant correlation between different parts of the measurement system. Final derivation of a structured equation model of the measurement system can be done the concept of entropy (Sect. 5.4.1).

2.5 Evaluation and Validation of a Measurement Method

Having designed and configured a measurement method as part of a measurement system it is necessary, before proceeding with full-scale measurement production, to evaluate and validate the method. 'Evaluation' means putting a number on the performance of a method (where performance metrics are typically those given in Sect. 2.4.4), while 'validation' means confirming that the method actually measures what is intended to be measured.

2.5.1 Design of Interlaboratory Experiment

Such evaluation and validation of a measurement method can be done as a fundamental investigation of a method (or a laboratory) in order to establish performance concerning accuracy. This will allow an assessment of whether the measurement method or laboratory has the potential to satisfy the demands placed on it as well as a validation or verification of an error budget formulated earlier.

An accuracy experiment according to ISO 5725-1 is often used for evaluation and validation of a measurement method. Such a method accuracy experiment is a kind of interlaboratory comparison (ILC) and is a valuable, collective exercise among a number of laboratories of known proficiency. An accuracy experiment is one in which the method, rather than the laboratories, is to be evaluated. Other applications of ILCs include evaluating laboratory proficiency and confirming claimed Calibration and Measurement Capabilities (CMC) as measures of the

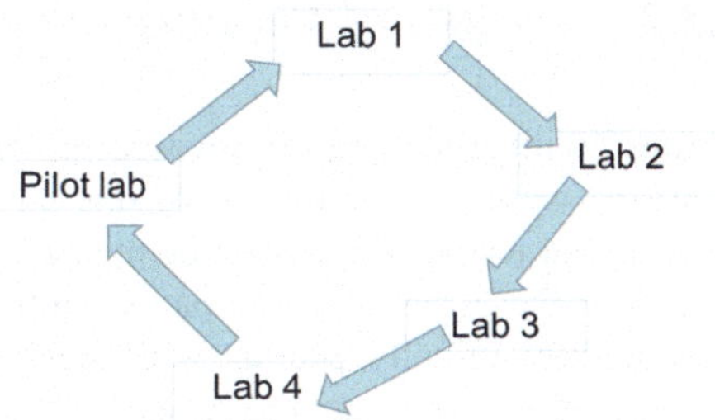

Fig. 2.8 Interlaboratory comparison (ILC)

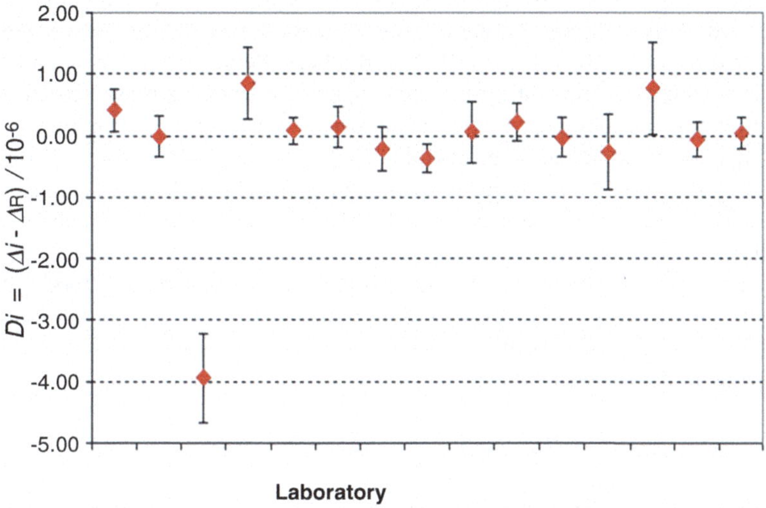

Fig. 2.9 Typical results (including uncertainty intervals, coverage factor, $k = 1$) of an interlaboratory comparison

metrological performance of the different actors (in cases where the method accuracy is known) (CIPM MRA 1999). Proposed new definitions of measurement units and their realisation are also evaluated and verified in ILCs.

A method accuracy experiment is in the form of an interlaboratory comparison (Fig. 2.8) in which typically test objects (preferably demonstrated in advance to be stable, even under transport) representing a number of levels, q, of the measurand are circulated by the pilot laboratory among a number, p, of laboratories whose proficiency has been determined earlier (in an interlaboratory comparison for proficiency testing (PT)). Apart from a simple loop (Fig. 2.8), there are several more elaborate schemes, such as star shapes, with which the travelling standards might be circulated by the pilot among the participants. The choice of scheme depends on factors such as the maximum time a travelling standard can be allowed to travel, etc.

2.5.2 Analysis of Variance in an ILC

Figure 2.9 illustrates some typical results from an ILC reported to the pilot laboratory from the various participants for one circulating object. It is not unusual, at least in an initial round of measurements, to observe significant differences between the results of the different participants, with deviations well outside the measurement uncertainties initially estimated by each participant individually. It is indeed the purpose of the experiment to reveal and subsequently remedy such discrepancies but note there are examples where the single ‘outlying’ result (such as shown in Fig. 2.9) turned out in the end to be the only correct result. A famous example of such ‘intellectual phase-locking’ among a group of laboratories was the measurement of the speed of light by Bergstrand (1952) and others (Petley 1985). A second classic example, graphically illustrating the challenges of estimating measurement uncertainty, was an interlaboratory experiment where the results of weighing a set of travelling mass standards among several of Europe’s leading laboratories differed by many times the quoted uncertainties, before one realised that the travelling standards themselves, being newly manufactured for the exercise, were the main source of instability.

An accuracy experiment can be described using a measurement model similar to that formulated (Eq. (2.1)). Instead of an individual[3] measurement system, in the case of an accuracy experiment one models the response output, y, of each of several participants in the experiment as:

$$y_{jik} = z_j + B_i + \varepsilon_{jik}; j = 1 \ldots q, i = 1 \ldots p_j, k = 1 \ldots n_{ij} \tag{2.12}$$

where the attribute (stimulus input) value, z, of the measurement object (circulating test object) can be distributed about a mean value $\bar{z} = \mu$ for each level (j). The other variables on the right-hand side of Eq. (2.12) denote variations distributed about a mean value of 0 (zero) for each element of the accuracy experiment (B for the laboratory, i, variation), and where ε_{jik} represents repeatability variations (ISO 5725).

A table, similar to Table 2.1, is established as a template for recording responses, y, and for ANOVA analysis of the interlaboratory comparison accuracy experiment. The columns in such a table will represent the different levels, j, while the rows are associated with each laboratory, i. From the analysis of the matrix of experimental results y_{jik}, usual statistical tools found in any statistical handbook (Montgomery 1996) can be deployed to calculate for instance s_L^2, an estimate of the between-laboratory variance; s_W^2, an estimate of the within-laboratory variance; s_r^2, repeatability (precision) variance, calculated as the arithmetic mean of s_W^2 and $s_R^2 = s_L^2 + s_r^2$, the reproducibility (precision) variance.

At this stage, several statistical methods can be employed to:

[3]The ILC set-up, including circulating object, pilot laboratory and other participants, with their respective measurement resources, can of course be regarded collectively as one measurement system.

- determine the scope of an accuracy experiment, for example, how many laboratories need participate, with statistical derived confidence levels for the trueness and precision of the method [ISO 5725-1] as an example of design of experiment (Sect. 2.1),
- identify and possibly eliminate outlying measurement data which might disproportionally affect the mean and standard deviation with the support of various statistical tests [ISO 5725-2],
- at the same time being aware of the risk of the above-mentioned 'intellectual phase-locking'.

2.5.3 Qualitative Accuracy Experiments

The measurement model, expressed in Eq. (2.12), of an accuracy experiment, similarly to the corresponding measurement model for a measurement system Eq. (2.1), is based on a number of assumptions: allowance has to be made for the effects of non-linearity, interferences, etc. as well as that the response may be ordinal, as is of interest in measurements in the social sciences, for instance.

Qualitative test methods also need to be evaluated and validated in an accuracy experiment as far as possible analogous to more quantitative methods. This is basically the same discussion of performance metrics and restitution for decision-making elements of a measurement system, as already given in Sects. 2.4.3 and 2.4.4, respectively.

Special tools are needed to formulate a measurement model analogous to Eq. (2.12) in the case where the test method produces responses on ordinal or nominal scales, since on such scales it is not sure that intervals on different parts of the scale are the same or even known. In such cases, the regular tools of statistics, even the simple calculation of means and standard deviations, cannot be assumed applicable in all cases.

It is clear there is no meaning in attempting to produce plots of the kind shown in Fig. 2.9 for the results of an interlaboratory comparison where the response of each measurement system is a performance metric of the kind probability of success, e.g. of making a correct identification of a chemical species. There are two basic reasons for not doing such plots (even though they are more frequently done than one would imagine). The response 'probability of success':

- is not on a quantitative scale, but usually on an ordinal scale of the counted fractions type (see Sect. 3.5). Distances (location and dispersion measures) on the y-axis will have no quantitative meaning,
- is a compound metric, consisting of a combination of the ability of the measurement system to make an identification and the level of difficulty posed by the task.

Approaches to calculating precision data from accuracy experiments for qualitative test methods as in Eq. (2.8) have been recently reported by among others Bashkansky et al. (2012) and Uhlig et al. (2013). The former authors attempted to characterise the accuracy of a measurement system in terms of the magnitudes of the various elements of a (binary) confusion matrix $\begin{pmatrix} 1-\alpha & \alpha \\ \beta & 1-\beta \end{pmatrix}$: Quoting from the recently proposed VIN (International vocabulary for nominal properties, (Nordin et al. 2018)): '3.9 *examination uncertainty*: fraction of examined values (3.5) that is different from a reference nominal property value (3.3) among all the examined values provided'. These decision risks are a good measure of precision, but do not capture the other component of accuracy, namely trueness, which is essential if one is to define metrological references for calibration purposes. Since the decision risks α and β have ordinal properties, it is not obvious that the usual tools of statistics can be used to describe them. A latent variable approach can circumvent some of this, as proposed by Akkerhuis et al. (2017), but their approach and that of Bashkansky et al. (2012) and Nordin et al. (2018) did not extend to include the all-important restitution process which would furnish an estimate of the sought-after measurand value separately from instrument ability (Sect. 2.4.4).

The Rasch invariant measurement theoretical approach is in our opinion an appropriate tool to handle both of the above bullet points in binary, nominal or ordinal test method accuracy experiments, as described both in the introductory chapter (Sect. 1.1), Sect. 2.4.5 and in detail later in Chap. 3.

2.5.4 Applications of Method Accuracy

Once determined in an accuracy experiment, the accuracy of a method can be used subsequently in diverse applications [ISO 5725-6]:

- Using accuracy in making a choice between two measurement methods,
- Product tests under conditions of repeatability and reproducibility,
- Process capability,
- SPC-diagrams,
- Trend analysis.

We will return to these applications later in this book (Sect. 6.3.3).

2.6 Verification

Together with validation—checking that one is measuring what was intended—the concept of verification, i.e. checking that the measurement system is reliable, is an essential part of metrological conformity assessment, to be described in more detail in Chap. 4.

Measurement *reliability*, according to Roach (2006), is a gauge of whether an outcome measure produces the same number each time a measurement instrument is administered. In using persons as measurement instruments, it is necessary to separate repeated measurements with one individual from measurement with different persons. There are several aspects to reliability; here we mention three of these:

1. Self-reporting: Reliability estimated from wording and interpretation,
2. Internal consistency of accessibility: Do all items in the outcome measure address the same underlying concept?
3. Rater performance: intra-rater consistency with repeats (reliability) and inter-rater consistency among different raters.

A reliability coefficient (R_z) for the item attributes, z, is calculated as:

$$\text{Reliability}, R_z = \frac{\text{True variance}}{\text{Observed variance}} = \frac{\mathrm{Var}(z)}{\mathrm{Var}(z')} = \frac{\mathrm{Var}(z') - \mathrm{Var}(\varepsilon_z)}{\mathrm{Var}(z')} \tag{2.13}$$

Once again, care has to be exercised when dealing with qualitative data (Sects. 2.4.4, 2.4.5 and 2.5.3) when evaluating this reliability coefficient (Eq. (2.13)) and analogous expressions, such as Cohen's kappa for rating judge reliability (Cohen 1960), since the various variance calculations are non-trivial on scales where exact distances are not known (Bashkansky and Gadrich 2012). Our recommendation is to apply a Rasch transformation to the measurement system response (P_{success}), according to the restitution process described in Sect. 2.4.5, where the ordinal, 'counted fraction' data is transformed onto a more quantitative scale.

2.7 Case Studies

2.7.1 Practical Working of a Measurement System

In the practical working of a measurement system, factors such as hardware, software, procedures and methods, human effort, etc., combine to cause variation among measurements of the same object that would not be present if the system were perfect.

To illustrate how this is tackled, we take up again the example of pre-packaged goods with the measurement specifications given in Table 2.3, when considering the appropriate characteristics of measurement systems for the task of testing product (Table 2.5).

An illustration of the measurement system, and the measurement process—including observation, calibration and restitution—for this example is shown in Fig. 2.10. (A more formal modelling will be presented in Chap. 5).

Table 2.5 Functional characteristics of appropriate measurement systems, based on test demands (Table 2.3)

For at least one of the chosen measurement systems (for the A non-functional or B functional product tests), give a description of the **functional** characteristics of the measurement system:	Your answers … *Coffee powder pre-packaged*
B Test of product **function**. Is the product 'right'? – Describe the chosen measurement system:	*The packet should contain as promised: correct amount powder - 500g.* *- Balance*
Describe how the measurement system is built up in terms of elements such as: sensor - signal conversion - signal conditioning - data conditioning, according to the instrument manufacturer or other specification.	*Mass => Balance (sensor => conversion => conditioning) => Reading => Appraiser (Sect. 2.4.2)*
Describe the characteristics (range, swing, non-linearity, sensitivity, etc) of the measurement system according to the instrument manufacturer or other specification.	*Range 0 - 1 kg; resolution 0.2g; sensitivity, $K = 1$ (Table 2.4)*
Describe test of these characteristics of the measurement instrument.	*Register output reading (O) of balance over a range of known input masses (I) to evaluate e.g. sensitivity K and bias b* $O = K \cdot I + N(I) + K_M I_M I + K_i I_i + b$ *(Eq. 2.4)*

2.7.2 *Man As a Measurement Instrument, Psychometry and Product Function*

Continuing the pre-packaged goods example, let us illustrate a measurement system appropriate for testing the functional requirements specified in Tables 2.3 and 2.6.

An illustration of the measurement system, and the measurement process—including observation, calibration and restitution—for this example is shown in Fig. 2.11. In the next chapter, we will consider in more depth the observation,

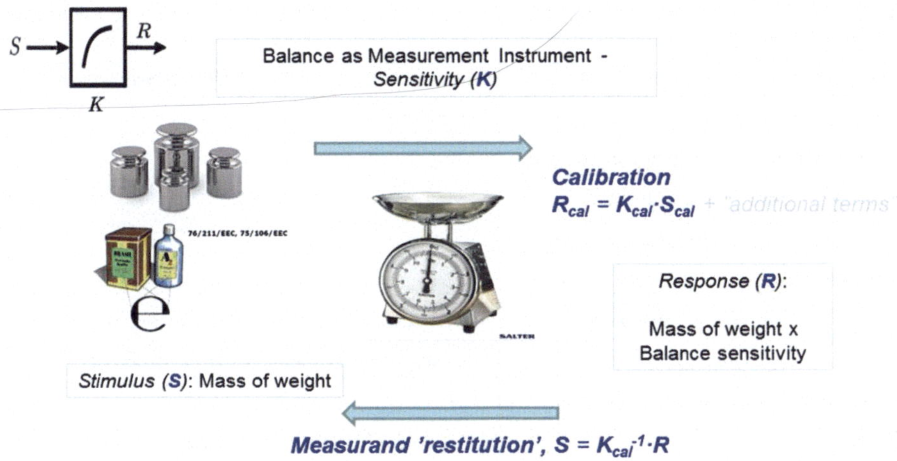

Fig. 2.10 Measurement system and process for functional characteristics (mass) of pre-packaged goods

calibration and restitution processes in cases such as this of functional characteristics of a more qualitative character. (A more formal modelling will be presented in Chap. 5).

Extending traditional metrological concepts in terms of a measurement from the physical sciences towards the social sciences, including measurements in psychometry and person-centred health, is a topic of current focus in the scientific literature. In work which fundamentally extends earlier measurement traditions—such as the systematic, representational and operational approaches (McGrane 2015)—guidelines are given (Pendrill 2014a, 2018) about how to describe measurements in for instance the psychological sciences in terms of a measurement system (ASTM 2012), as traditionally done in the physical sciences.

Table 2.6 Functional characteristics of appropriate measurement system, based in test demands (Table 2.3)

For each type of test (A non-functional; B functional) below, choose a measurement system than can be used to test product – Please attach a specification sheet or other description of each measurement system.	Your answers .... *Coffee powder pre-packaged*
B Test of product **function**. Is the product 'right'? – Describe the chosen measurement system:	*B. function. One experiences a certain 'prestige' when serving this mark of coffee:* *Measure of perceived 'prestige': 80%, at least 60% (customer survey)* - *Man as measurement instrument*
Describe how the measurement system is built up in terms of elements such as: sensor - signal conversion - signal conditioning - data conditioning, according to the instrument manufacturer or other specification.	***Measurement object***: *samples of pre-packaged coffee packets; quality characteristics: quality associated with prestige* ***Measurement instrument***: *Panel of consumers; quality characteristics: leniency. Perceived prestige response of each panellist to packet quality registered with a questionnaire with 10 items, and each item graded with 5 categories. Decision-making accuracy in terms of 'correctness' of categorisation,* $P_{success}$*., from* $\theta - \delta = \log\left(\frac{P_{success}}{1-P_{success}}\right)$ Eq. (1.1) or Eq. (2.11) (Sects. 2.4.4 and 2.4.5)

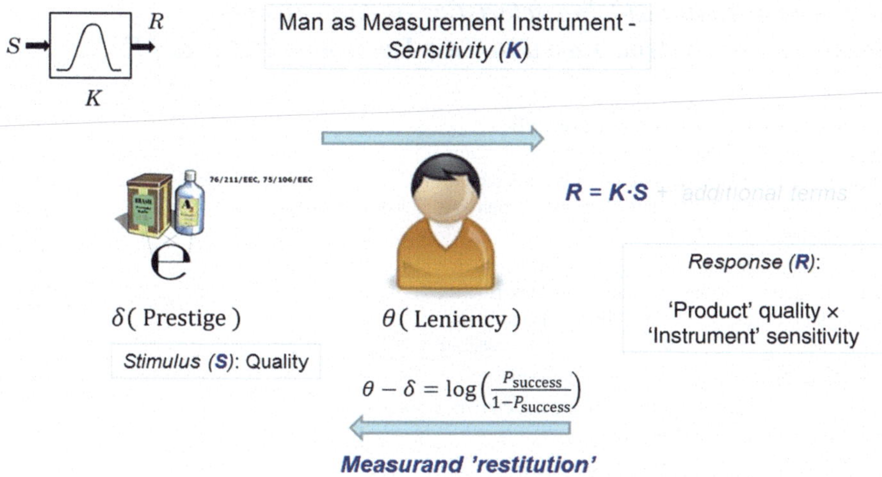

Fig. 2.11 Measurement system and process for functional characteristics (prestige) of pre-packaged goods

Exercise 2 Definition of Measurement Problem

E2.1 Demands on Measurement System and Methods, Based on Product Demands §E1.2:

	Your answers. . . . .
For as many as possible of each product test (A, B, C, D) under §E1.3, specify demands on measurement system and methods:	–
– Minimum production capability ($C_{p,\min}$)?	–
– Minimum measurement capability ($C_{m,\min}$)?	–
– Maximum permissible measurement uncertainty (*MPU*)?	–
– Maximum permissible measurement (*MPE of instrument/system*)?	–
Other demands:	–

E2.2 Non-functional Characteristics of Appropriate Measurement System, Based in Test Demands §E1.3 & §E2.1:

For each type of test (A non-functional; B functional) below, choose a measurement system than can be used to test product—Please attach a specification sheet or other description of each measurement system.	Your answers ....
A Test of product *non-functional characteristics*. Does the product work 'correctly'?—Describe the functional aspects of the chosen measurement system:	–
– Describe the non-*functional* characteristics	–
– Describe test of these functional characteristics of the measurement instrument	–
– What does each instrument test cost?	–
B Test of product *function*. Is the product 'right'?—Describe the chosen measurement system:	–
– Describe the *functional* characteristics (range, swing, non-linearity, sensitivity, etc.) of the measurement system according to the instrument manufacturer or other specification	–
– Describe test of these functional characteristics of the measurement instrument. Evaluate: $O = K{\cdot}I + N(I) + K_M{\cdot}I_M{\cdot}I + K_i{\cdot}I_i + b$	–
– What does each instrument test cost?	–
Others:	–

E2.3 Functional Characteristics of Appropriate Measurement Systems, Based on Test Demands §E1.3 & §E2.1:

For at least one of the chosen measurement systems (for the A non-functional or B functional product tests), give a description of the *functional* characteristics of the measurement system:	Your answers ....
A Test of product *non-functional characteristics*. Does the product work 'correctly'?—Describe the chosen measurement system:	–
– Describe how the measurement system is built up in terms of elements such as: sensor—signal conversion—signal conditioning—data conditioning, according to the instrument manufacturer or other specification	–

(continued)

For at least one of the chosen measurement systems (for the A non-functional or B functional product tests), give a description of the *functional* characteristics of the measurement system:	Your answers …………………………………………
– Model how errors in measurement signals are propagated through the measurement instrument. Evaluate: $O = K{\cdot}I + N(I) + K_M{\cdot}I_M{\cdot}I + K_i{\cdot}I_i + b$ for each element of each measurement instrument element and for the complete measurement system	–
– What does each instrument test cost?	–
Others:	–
B Test of product *function*. Is the product 'right'?—Describe the chosen measurement system:	–
– Describe how the measurement system is built up in terms of elements such as: sensor—signal conversion—signal conditioning—data conditioning, according to the instrument manufacturer or other specification	–
– Model how errors in measurement signals are propagated through the measurement instrument. Evaluate: $O = K{\cdot}I + N(I) + K_M{\cdot}I_M{\cdot}I + K_i{\cdot}I_i + b$ for each element of each measurement instrument element and for the complete measurement system	–
– What does each instrument test cost?	–

References

AIAG, Measurement systems analysis reference manual, in *Chrysler, Ford, General Motors Supplier Quality Requirements Task Force*, (Automotive Industry Action Group, Michigan, 2002)

T. Akkerhuis, J. de Mast, T. Erdmann, The statistical evaluation of binary test without gold standard: robustness of latent variable approaches. Measurement **95**, 473–479 (2017). https://doi.org/10.1016/j.measurement.2016.10.043

ASTM, Standard Guide for Measurement Systems Analysis (MSA) *E2782*. (2012), https://doi.org/10.1520/E2782-11

E. Bashkansky, T. Gadrich, Mathematical and computational aspects of treatment ordinal results, in *AMCTM IV, Series on Advances in Mathematics for Applied Sciences*, vol. 84, (World Scientific Publishing Co. Pte. Ltd, Singapore, 2012). ISBN-10 981-4397-94-6

E. Bashkansky, T. Gadrich, I. Kuselman, Interlaboratory comparison of test results of an ordinal or nominal binary property: analysis of variation. Accred. Qual. Assur. **17**, 239–243 (2012)

J.P. Bentley, *Principles of Measurement Systems*, 4th edn. (Pearson Education Limited, London, 2005)

B. Berglund, G. B. Rossi, J. T. Townsend, L. R. Pendrill (eds.), *Measurement With Persons: Theory, Methods, and Implementation Areas* (Psychology Press, Scientific Psychology Series, London, 2011). Published: December 2011 ISBN: 978-1-84872-939-1

E. Bergstrand, *Recent Developments and Techniques in the Maintenance of Standards* (HMSO, London, 1952). Ann. Fr. Chronom. **11**, 97. ibid. 1957

CIPM MRA, International Equivalence of Measurements: The CIPM MRA. (1999), https://www.bipm.org/en/cipm-mra/

J.A. Cohen, A coefficient of agreement for nominal scales. Educ Psychol. Meas. **20**, 37–46 (1960)

EU Commission, **Directive 2014/32/EU** of the European Parliament and of the Council of 26 February 2014 on the Harmonisation of the Laws of the Member States Relating to the Making Available on the Market of Measuring Instruments (2014)

EU Commission, Conformity Assessment. (2018), https://ec.europa.eu/growth/single-market/goods/building-blocks/conformity-assessment_en

A. Gillespie, F. Cornish, Intersubjectivity: towards a dialogical analysis. J. Theory Soc. Behav. **40**, 19–46 (2010)

J.P. Guilford, *Psychometric Methods* (McGraw-Hill, Inc, New York, 1936), pp. 1–19

S.M. Humphry, The role of the unit in physics and psychometrics. Meas. Interdisciplinary Res. Perspect. **9**(1), 1–24 (2011)

ISO 10012:2003 *Measurement management systems -- Requirements for measurement processes and measuring equipment*, ISO

ISO 11095:1996 Linear calibration using reference materials

ISO 5725:1995 *Accuracy – trueness and precision*, in 6 parts

ISO 9001:2015 Quality management systems — Requirements, https://www.iso.org/obp/ui/#iso:std:iso:9001:ed-5:v1:en

ISO Guide 33:2015 Reference materials -- Good practice in using reference materials

ISO/IEC 17025:2017 General requirements for the competence of testing and calibration laboratories, https://www.iso.org/standard/66912.html

G. Iverson, R. Luce, The representational measurement approach to psychophysical and judgmental problems, in *Measurement, Judgment, and Decision Making*, (Academic Press, Cambridge, 1998)

JCGM 100:2008 Evaluation of measurement data – Guide to the expression of uncertainty in measurement (GUM 1995 with minor corrections) in *Joint Committee on Guides in Metrology (JCGM)*

JCGM 106:2012, “Evaluation of measurement data – The role of measurement uncertainty in Conformity Assessment”, in *Joint Committee on Guides in Metrology (JCGM)*

JCGM 200:2012 International vocabulary of metrology—basic and general concepts and associated terms (VIM 3rd edition) (JCGM 200:2008 with minor corrections) *WG2 Joint Committee on Guides in Metrology (JCGM)* (Sevrès: BIPM)

H. Källgren and L.R. Pendrill, Exhaust gas analysers and optimised sampling, uncertainties and costs, Accreditation and Quality Assurance – Journal for Quality, Reliability and Comparability in Chemical Measurement. **11**, 496–505, (2006) https://doi.org/10.1007/s00769-006-0163-3

A. Kohlhase, M. Kohlhase, Semantic knowledge management for education. Proc. IEEE **96**, 970–989 (2008)

J.M. Linacre, Optimizing rating scale category effectiveness. J. Appl. Meas. **3**(1), 85–106 (2002)

P. Loftus, S. Giudice, Relevance of methods and standards for the assessment of measurement system performance in a high-value manufacturing industry. Metrologia **51**, S219–S227 (2014)

J. McGrane, Stevens' forgotten crossroads: the divergent measurement traditions in the physical and psychological sciences from the mid-twentieth century. Front. Psychol. Hypothesis Theory **6**, 1–8 (2015). https://doi.org/10.3389/fpsyg.2015.00431. art. 431

D.C. Montgomery, *Introduction to Statistical Quality Control* (Wiley, Hoboken, 1996). ISBN: 0-471-30353-4

G. Nilsson, private communication (1995)

G Nordin, R Dybkaer, U Forsum, X Fuentes-Arderiu and F Pontet, Vocabulary on nominal property, examination, and related concepts for clinical laboratory sciences (IFCC-IUPAC recommendations 2017), Pure Appl. Chem. **90**(5): 913– 935, (2018)

I. Partchev, *A visual guide to item response theory* (Friedrich-Schiller-Universität Jena, Jena, 2004). https://www.coursehero.com/file/28232270/Partchev-VisualIRTpdf/

L R Pendrill, Meeting future needs for metrological traceability – a physicist's view – Accred. Qual. Assur. J. Qual. Reliab. Comparability Chem. Meas., **10**, 133–9. (2005). http://www.springerlink.com/content/0dn6x90cmr8hq3v4/?p=2338bc01ade44a208a2d8fb148ecd37api

L.R. Pendrill, "El ser humano como instrument de medida". *e-medida*. (2014a)

L.R. Pendrill, Man as a measurement instrument. NCSLI Meas. J. Meas. Sci. **9**, 24–35 (2014b)

L.R. Pendrill, Using measurement uncertainty in decision-making & conformity assessment. Metrologia **51**, S206 (2014c)

L.R. Pendrill, Assuring measurement quality in person-centred healthcare. Measurement Science & Technology **29**, 034003 (2018). https://doi.org/10.1088/1361-6501/aa9cd2. special issue Metrologie 2017

B.W. Petley, *The Fundamental Physical Constants and the Frontier of Measurement* (Adam Hilger Ltd, Bristol, 1985). ISBN 0-85274-427-7

K.E. Roach, Measurement of health outcomes: reliability, validity and responsiveness. J. Prosthet. Orthot. **18**, 8 (2006)

G.B. Rossi, *Measurement and Probability – A Probabilistic Theory of Measurement with Applications*, Springer Series in Measurement Science and Technology (Springer, Dordrecht, 2014). https://doi.org/10.1007/978-94-017-8825-0

G.B. Rossi, F. Crenna, "Toward a formal theory of the measuring system", IMEKO2016 TC1-TC7-TC13. J. Phys. Conf. Ser. **772**, 012010 (2016). https://doi.org/10.1088/1742-6596/772/1/012010

M. Thompson, T. Fearn, What exactly is fitness for purpose in analytical measurement? Analyst **121**, 275–278 (1996)

S. Uhlig, S. Krügener, P. Gowik, A new profile likelihood confidence interval for the mean probability of detection in collaborative studies of binary test methods. Accred. Qual. Assur. **18**, 367–372 (2013)

M. Wenemark, *Questionnaire Methodology with a Focus on the Respondent (in Swedish)* (Studentlitteratur, Lund, 2017). ISBN 978-91-44-09641-4

Chapter 3
Ensuring Traceability

This chapter deals with traceability and comparability: the first of the two major hallmarks of metrology (quality-assured measurement). The second hallmark—uncertainty—is covered in the final chapters of the book.

In Fig. 3.1 is recounted a parable illustrating the concept of measurement comparability, in this case of the physical quantity of time, but the meaning should be universal.

A retired sea-captain (a) on an island takes his time from the watchmaker (b) in town only to find out that the watchmaker uses the sea-captain's cannon shots at 12 noon each day to set his own clocks! (Attributed to Harrison (MIT) by Petley 1985). Examples of this kind of circular traceability in measurement are more common than one would hope. The parable captures the essence of the concept of trueness, that is, what defines being 'on target' when making repeated measurements in the 'bull's eye' illustration of Fig. 2.2?

Reliable measurement results are important in almost every aspect of daily life. Comparability of entity properties is critical to global trade between producers and consumers; provides for the interoperability and exchangeability of parts and systems in complex industrial products; is essential in the synchronisation of signals in communication systems and provides a fair and quality-assured basis for measurement in the environmental, pharmaceutical and other chemical sectors. Metrological traceability through calibration enables the measurement comparability needed, in one form or another, to ensure entity comparability in any of the many fields mentioned in the first lines of Chap. 1.

Despite its importance, international consensus about traceability of measurement results—both conceptually and in implementation—has yet to be achieved in every field. Ever-increasing demands for comparability of measurement results needed for sustainable development in the widest sense require a common understanding of the basic concepts of traceability of measurement results at the global level, in both traditional as well as new areas of technology and societal concern.

The present chapter attempts to reach such a consensus by considering in depth the concept of traceability, in terms of calibration, measurement units and standards

L. Pendrill, *Quality Assured Measurement*, Springer Series in Measurement Science and Technology, https://doi.org/10.1007/978-3-030-28695-8_3

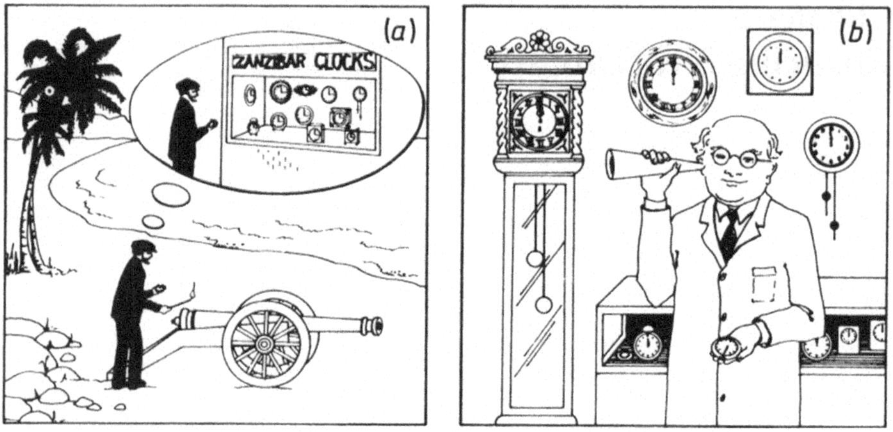

Fig. 3.1 Zanzibar effect (Reproduced with permission Petley 1985)

(etalons), symmetry, conservation laws and entropy, in a presentation founded on quantity calculus. While historically physics has been the main arena in which these concepts have been developed, it is now timely to take a broader view encompassing even the social sciences, guided by philosophical considerations and even politics. At the same time as the International System of Units is under revision, with more emphasis on the fundamental constants of physics in the various unit definitions, there is some fundamental re-appraisal needed to extend traceability to cover even the less quantitative properties typical of measurement in the social sciences and elsewhere.

3.1 Quantity Calculus and Calibration

What use can be had of quantity calculus when seeking measurement comparability needed, in one form or another, to ensure entity comparability in any of the many fields of application?

With much of measurement developed over many years in the context of physics and engineering, there has been a long-standing connexion with applied mathematics: 'The object of measurement is to enable the powerful weapon of mathematical analysis to be applied to the subject matter of science' according to Campbell (1920). The renowned theoretical physicist Feynman (2013) went as far as to state: 'Experiment is the *sole judge* of scientific "truth."'

In developing theory in quantum mechanics almost one hundred years ago, Dirac [§5, p.15 1992) wrote: 'In an application of the theory one would be given certain physical information, which one would proceed to express by equations between the mathematical quantities. One would then deduce new equations with the help of the axioms and rules of manipulation and would conclude by interpreting these new equations as physical conditions. The justification for the whole scheme depends, apart from internal consistency, on the agreement of the final results with experiment'.

Quantity calculus, as developed over much of the twentieth century, is described echoing Dirac from the metrological point of view as: 'mathematical relations between quantities in a given system of quantities, independent of measurement units' (JCGM 200: 2012, VIM §1.22). Quantity calculus, according to the review of de Boer (1994/5), has its roots in the mathematical formalism of theoretical physics dating from the pioneering work of Wallot (1926). At the same time, it cannot be said that there exists yet a general consensus about an axiomatic foundation for quantity calculus (de Boer 1994/5; Raposo 2016). This incompleteness is becoming of increasing concern as interest in quality-assured measurement spreads to new sectors such as the social sciences.

3.1.1 Quantity Concepts

In defining quantity calculus, an hierarchy of concepts is traditionally established in the order: Kind of quantity; Quantity; Value of quantity. Fleischmann (1960) gave some examples of the concepts at the different hierarchy levels (my translation from the German):

- 'For each single feature, every kind of quantity can have an infinite number of quantities (characteristic quantities) of the feature.
- When the quantity is not specific, one refers to a 'Quantity' or 'General quantity'. For example: the quantity electric voltage V can take the value 1 V or the value 20 V, etc.
- An 'entity quantity', in the terminology of Fleischmann (1960),[1] is the case where a general quantity can be associated with an entity (object). For example, the effective voltage and maximum voltage (for sinusoidal alternating current) are distinct entity quantities.
- Quantity values are specific values of quantities. For instance in the area of electrical voltage, one can have quantity values 5 mV, 30 V, etc. Examples of quantity values which belong to different kinds of quantity are 10 s and 3 cm. Quantity values are associated with a definite quantity.'

The superordinate concept of 'kind of quantity' in quantity calculus (as in Fleischmann 1960) is used to collect together quantities by kind or character. Quantities themselves and relations between them, that is, in the 'entity (or product) space' (Chap. 1), irrespective of whether they are measured or not, come next in the conceptual hierarchy. Thereafter, aspects in the measurement space

[1] German: '*Sachgrösse* = Objektgrösse (mit Objektbindung behaftete Grösse). Sie hat Quantität, die aber unbestimmt bleibt, sie hat Sachbezug (Objektbezug)' Fleischmann (1960).

(Chap. 2), such as quantity values, can be introduced subsequently, as required (Sect. 3.2).

Which of these various groupings are chosen depends on what is needed and meaningful in each field of application when in general communicating measurement information.

In full generality, relations in quantity calculus can be formed between different kinds of quantity and quantities in a more abstract sense. Consideration of 'general' quantities is much the domain of the physicist (Pendrill 2005) where relations (laws of Nature) among such quantities (which also give the corresponding relations among the measurement units associated with them—see Sect. 3.2) are fundamentally and universally applicable, *irrespective* of particular objects (as for instance in Newtonian mechanics as applied to all bodies, from microscopic and cosmological scales). Measurement is a necessary action in physics, but the main interest is in understanding the universe.

Physical quantities possess some remarkable properties which enable correspondingly remarkable possibilities for metrologically traceable measurements: not only can the results of measurements of a particular quantity be compared, but also measurements of *different quantities* can show a degree of comparability (Sect. 3.6.1). Relations between physical quantities can be expressed mathematically with equations that express laws of Nature or define new quantities.

Quantity calculus can be used to highlight the interesting distinction between:

- a physical law, e.g. force $F = m \cdot a$ (Newton's second law, if the mass, m, is constant) relating different quantities is universal; applicable at all scales from the microscopic to the cosmological,
- those indirect measurements where only an empirical 'recipe' is used, e.g. an engineering expression relating a number of different properties for a particular object, which has local validity but only limited universality. An example is an expression for the volume, V, of combustion chamber above a combustion engine piston $V = \varepsilon\ V_k$, where V_k – volume of combustion chamber when piston is in upmost position; ε – compression ratio (Kogan 2014).

The remarkable properties in physics are of course not necessarily shared by quantities in other disciplines (Pendrill 2005; Nelson 2015), which is a key issue of course in the context of the present book, aiming to give a unified presentation of quality-assured measurement across the social and physical sciences. In the last column of Table 3.1, an example—for the unit of length—is given to illustrate the various ways of expressing the unit at each level of the quantity calculus hierarchy.

Table 3.1 Comparing concepts in information theory and quantity calculus

Quantity calculus	Information theory	Examples: SI unit of length, metre, symbol m
Nature of quantity	*Effectiveness*—'changing conduct': relationship between signs of communication and actively 'improving' the entities they stand for (Weinberger 2003)	Quality of cloth products sold by metre length
Kind of quantity	*Pragmatic*—'utility': relationship between signs of communication and their utility (value, impact)	Length of cloth costing 10€/m
Entity quantity	Object quantity (object-bound quantity). It has quantity, but remains indefinite; it has reference to reality (object reference) (Fleischmann 1960)	Length of cloth
Quantity	*Semantic*—'meaning': relationship between signs of communication and entities they stand for	Distance travelled by light in $1/c$ seconds ('explicit unit' definition, CGPM 2019)
Value of quantity	*Syntax*—'signs': relationship among signs of communication such as numbers	Defined by taking fixed numerical value of speed of light in vacuum c to be 299 792 458 when expressed in unit m/s, where the second is defined in terms of $\Delta\nu_{Cs}$ ('explicit constant' definition, CGPM 2019)

3.1.2 *Introducing Measurement and Calibration. Separating Object and Instrument. Restitution*

At the lowest level of the hierarchy of concepts of quantity calculus, mathematical relations can be formulated specifically between quantity values, firmly in the realm of measurement.

In general, the value of an item attribute δ (e.g. a level of challenge of a particular task or the mass of a weight) differs from the 'true' δ', by an error ε_δ:

$$\delta = \delta' + \varepsilon_\delta \tag{3.1}$$

For comparability, it is necessary to communicate information about the extent to which the item attribute value is in error. As explained in Sect. 2.1, it is necessary (and challenging) to distinguish between actual product value and an apparent value distorted by limited measurement reliability.

The calibration process—that is, determining the error (with respect to the 'true value' depicted as the bull's eye at the centre of the target shown in Fig. 2.2)—often identifies the attribute of a particular object (called a measurement standard or etalon) as a known reference with which other attributes (and their errors) can be referenced.

The international metrology vocabulary defines (JCGM VIM 200:2012):

5.1 *measurement standard* etalon

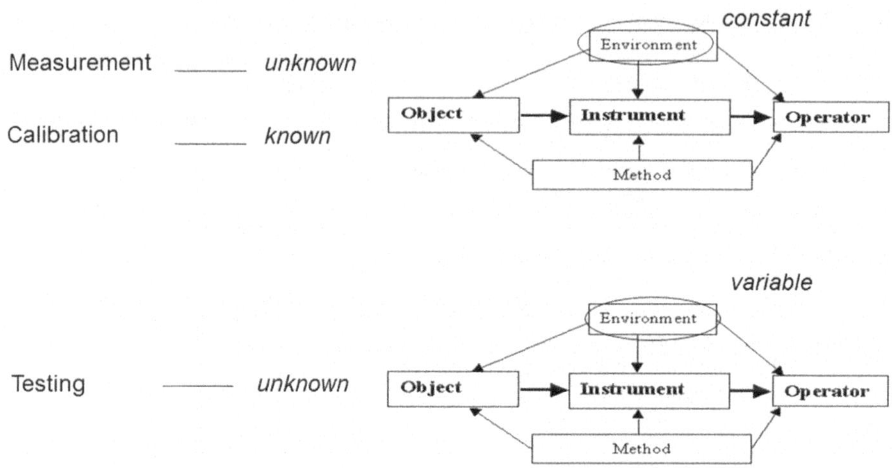

Fig. 3.2 Measurement, calibration and testing under different conditions of a measurement system

> realization of the definition of a given quantity, with stated quantity value and associated measurement uncertainty, used as a reference
>
> NOTE 1 A 'realization of the definition of a given quantity' can be provided by a measuring system, a material measure, or a reference material.
>
> NOTE 2 A measurement standard is frequently used as a reference in establishing measured quantity values and associated measurement uncertainties for other quantities of the same kind, thereby establishing metrological traceability through calibration of other measurement standards, measuring instruments, or measuring systems.

Calibration of a measurement system for instance consists of applying a stimulus of known value (from the standard) to the input of the measurement system, and then determining the extent to which measurement error can be associated with every element of the measurement system (Fig. 2.5).

The process of calibration is illustrated in Fig. 3.2 in comparison with the related but distinct operations of testing a measurement system as well as making measurements with a (calibrated) measurement system to determine an unknown quantity.

Calibration is defined in the international vocabulary (VIM §2.39) as:

> Operation that, under specified conditions, in a first step, establishes a relation between the quantity values with measurement uncertainties provided by measurement standards and corresponding indications with associated measurement uncertainties and, in a second step, uses this information to establish a relation for obtaining a measurement result from an indication.

In terms of measurement system analysis, it is important to distinguish between the quantity value which is the measured response of the instrument to a known stimulus, and the value of the entity as estimated with the process of restitution. We interpret the 'measurement result from an indication' in the VIM calibration definition as referring to the latter.

A calibrated measurement system is of course a pre-requisite in a usual measurement situation where one uses the system to determine the unknown stimulus from the measurement object. The restitution process which converts the system response into the stimulus value requires a known sensitivity of the measurement instrument (Sect. 2.4.5). In the case where the item (entity) is acting as a calibration standard (etalon), then the error ε_δ (Eq. (3.1)) is known (from a previous calibration process). Observing the response of the measurement system being calibrated and tested with the known stimulus allows determination of in principle all characteristics associated with the measurement system (listed in Table 2.4). Among the most frequently determined characteristics when calibrating a measurement system are the sensitivity, K, of the instrument and the bias, b. For instance, $b_{\mathrm{cal}} = b - \varepsilon_\delta$. If both K and b are known from calibration and can be assumed to be stable enough[2] that they remain substantially unchanged in value on later usage of the calibrated measurement system, then the process of restitution yields an estimate of an unknown stimulus in the simplest case with the expression

$$S = \frac{R - b_{\mathrm{cal}}}{K_{\mathrm{cal}}} \tag{2.9}$$

If the ultimate aim of a measurement is not only to determine an error, but to make a decision of conformity for an entity, then restitution will yield estimates, δ, of the respective properties of interest—for example, measurements in the social sciences such as of the difficulty of a task; the quality of a service or beauty of the portrait. As described earlier (Sect. 2.4.5), restitution for categorical responses (e.g. of a human being's response as an 'instrument' to a stimulus) is made in terms of P_{success}, that is, the probability of making a correct categorisation and the Rasch measurement model, mentioned in Sect. 1.2.3 and Eq. (1.1) can be applied:

$$S = z = \theta - \delta = \log\left[\frac{P_{\mathrm{success}}}{1 - P_{\mathrm{success}}}\right] \tag{1.1}$$

In contrast to physical measurement systems, the measurand (quantity to be assessed in the social sciences, such as the degree of quality of care, δ)—which is the stimulus S of the measurement object characteristics—is estimated through restitution of the measurement system response R of instrument (sensitivity K) in terms of a performance metric (how well the task is performed or how well the quality of a service is rated) using Rasch invariant measurement theory, rather than an inversion merely based on measurement error.

Measurement is a 'concatenation of observation and restitution' (as recalled by Bentley 2004; Sommer and Siebert 2006; Rossi 2014).

A full picture of the measurement process when Man acts as a measurement instrument can be given, as in Fig. 3.3, presenting the process, step by step, from the observed indication (a performance metric, e.g. probability of success, P_{success}, of

[2]Chapter 5 contains a description of how to test these assumptions.

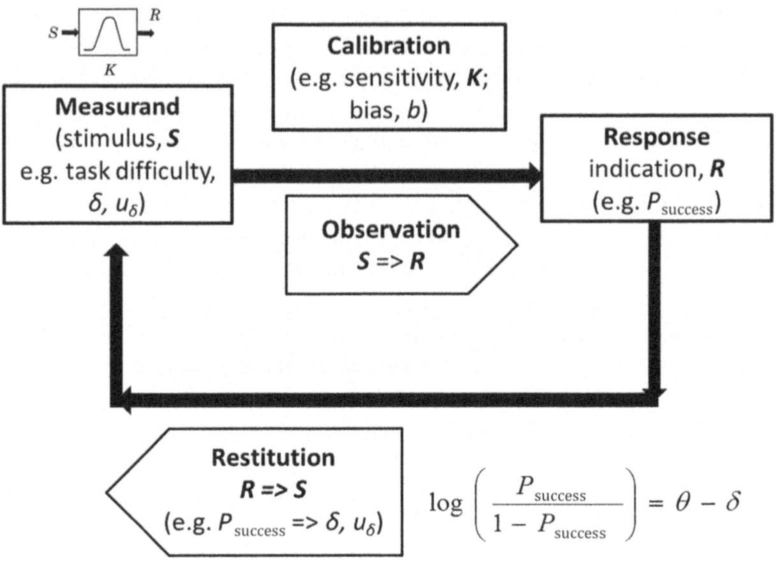

$$\log\left(\frac{P_{\text{success}}}{1 - P_{\text{success}}}\right) = \theta - \delta$$

Fig. 3.3 Observation and restitution for performance metrics (adapted from Pendrill 2018)

achieving a task), and restitution with Rasch Measurement Theory, through to the measurand (e.g. task difficulty) in a form suitable for metrological quality assurance.

Irrespective of whether measurements are made in the physical or social sciences, a fundamental requirement for establishing metrological standards (etalons) for calibration (be it of other etalons or measurement systems) is that a separation can be made in the response of the measurement system between the stimulus value—that of the measurand associated with the measurement object—and the characteristics of the measurement system—particularly the sensitivity and eventual bias of the measurement instrument. Without that separation, there is no way of reliably performing the restitution process nor establishing the all-important metrological aspects of traceability and comparability.

In connexion with the introduction of a revised SI (Sect. 3.6.1), the exact terminology to be used to describe quantities, quantity values, and units has been debated in a number of international committees as well as in the research literature (Kogan 2014; Mari et al. 2018). In a first case study (Sect. 3.6.1) at the end of this chapter, we consider how quantum calculus can be invoked to make clear the conceptual difference between quantity and quantity value. In a second case study (Sect. 3.6.2), establishment of metrological standards for calibration in the social sciences (and other 'qualitative' situations) will be considered.

3.2 Units and Symmetry, Conservation Laws and Minimum Entropy

Communicating information about the extent with which the item attribute is in error can be seen as a description of metrological comparability in a purely mathematical sense as in Eq. (3.1). But measurement is not only mathematics and in the present section we consider the additional aspects which guide us when choosing among the various groupings of quantity calculus (Sect. 3.1.1). In particular, what kind of measurement information is being communicated and what is the 'meaning' behind an error ε_δ?

Which grouping of quantity calculus concepts one chooses depends on what is needed in each field of application when communicating measurement information in general. Roberts (1985), in his introduction to measurement with applications in the social sciences, illustrates this with a number of statements (or messages) which appear either meaningful or meaningless (Table 3.2).

Two issues about measurement are raised by Roberts' (1985) examples:

- Measurement has much to do with providing clear, meaningful messages with which to communicate information about what (entity, quality characteristic…) is being measured.
- Measurement units are a key concept in enabling efficient communication.

In this section, these and related concepts such as entropy, symmetry, conservation will be presented which in different ways relate to the efficient and meaningful communication of measurement information. The introduction of meaningfulness of empirical scalings and analyses based on them, while elementary, has not been without statistical controversy where some statisticians perceive the approach as too prescriptive (Velleman and Wilkinson 1993). Further discussion of quantitative and qualitative scales of measurement is given in Sect. 3.5.

Table 3.2 Meaningful messages about measurement

Number of cans of corn in local supermarket at closing time yesterday was at least 10	Meaningful
One can of corn weighs at least 10	Meaningless
One can of corn weighs twice as much as a second can	Meaningful
Temperature of one can of corn at closing time yesterday was twice as much as that of a second can	Meaningless

Adapted from Roberts 1985

3.2.1 Meaningful Messages and Communicating Measurement Information

Regarding measurement as a particular kind of information was briefly eluded to earlier in this book (Sect. 2.2.3) when introducing measurement system analysis (MSA) in terms of a faithful description of the observation process, that is, how measurement information is transmitted from the measurement object, via an instrument, to an observer. Apart from this local communication, it is of course in many cases also required to communicate measurement information as globally—across distances and among different persons—as needed, according to what is meaningful in each field of application (listed at the start of Chap. 1). In principle, the manner of formulating measurement information will depend on which level of the quantity calculus hierarchy one is at (Table 3.1).

When considering a suitable choice of types of quantities in quantity calculus in view of what kind of information is considered important to communicate, Emerson (2008) for instance argues that 'kind of quantity' (Sect. 3.1.1) has to be complemented by adding the distinct concept of 'nature of quantity', and gives examples of two quantities of the same kind, but which are 'meaningless' to compare:

- 'the length of the distance between two rail termini has a different nature to the length of the track gauge of the same railway,
- the height of tide at London Bridge which is the same kind of quantity, but of a different nature, to the height of tide at Washington Bridge'.

Other examples are given in Table 3.1. Emerson (2008) takes such examples of the comparability of kinds of quantities as illustrations of extensive quantities which can be added but only if their 'datum values are not associated with different and immovable places or times'.

Similar discussions may be found in infology (Langefors 1993), where 'information' is tied to the relevance of the knowledge to the decision to be made. Knowledge that has no bearing on the problem at hand will not reduce the uncertainty associated with the problem, and will not be recognised as information. A minimum information element would contain at least three parts: one part referring to the entity informed about; another part which refers to the property of the entity and a third, a 'locator', such as a time reference part in the sentence. For example: 'The temperature of container C was 65 degrees on April 15, 1992 in room X'. In modern information terminology, one could refer to these as conditions for maintaining the integrity of a message on transmission through a communication system—so-called semantic interoperability (Marco-Ruiz et al. 2016 give examples from health informatics). At the pinnacle of the quantity calculus hierarchy Weinberger (2003) identifies the concept of information *effectiveness*, meaning 'changing conduct' in terms of relations between signs of communication having to do with actively 'improving' the entities they stand for.

Transmission of measurement information is of course a specific case of communication with an information system (Pendrill 2011). Be it with a local measurement system (as depicted in Fig. 2.4) or more globally across distances and among different persons, the 'transmitter' (event 'B'), 'information channel' (signal 'B' => 'A') and 'receiver' (message 'A') of a classical information system in the case of a measurement system correspond, respectively, to the measurement object, measurement instrument and observer. The amount of information transmitted from the measurement object to the observer can range from a simple signal through to increasingly 'meaningful' messages, as is captured in four levels of increasing richness in information theory (Weaver and Shannon 1963; Klir and Folger 1988) as given in the first column of Table 3.1. Depending on what kind of meaning is to be communicated, the kind of (measurement) information will fall into one or other of the extended quantity calculus hierarchy (second column of Table 3.1).

3.2.2 Units, Words and Invariance

A useful point of departure for introducing units in measurement is to recognise the analogous role of words in making efficient communication with language. Metrological traceability enables the measurement comparability needed, and calibration (Sect. 3.1.2) involves tracing the measurement standards which, so to speak, embody defined and 'recognisable' or 'meaningful' amounts of the measurement unit.

A classic example illustrating the different information content of the following three messages consisting of an equal number of digits (or bits):

'100110001100'
'agurjerhjjkl'
'this message'

It is obvious to everyone who understands English that the third message conveys more information than the two other messages. In the next section a connexion will be made between the amount of information conveyed by a message and informational entropy, where the latter is a measure of the degree of 'order' in broad meaning in the message contents.

The international metrology vocabulary [VIM §1.9] defines *measurement unit* as a:

> real scalar quantity, defined and adopted by convention, with which any other quantity of the same kind can be compared to express the ratio of the two quantities as a number.

Because of the key role played by measurement units, there have to be both clear definitions of each unit as well descriptions of how each unit is 'realised'. To be of any practical use, units do not only have to be defined, but they also have to be realized physically for dissemination. A variety of experiments may be used to realise the definitions—called 'mise en pratique'. Current definitions of the measurement units of the International System (SI) can be found in the SI brochure (CGPM 2019).

The original text by Maxwell referring to measurement units with the expression Eq. (3.2):

$$Q = \{Q\} \cdot [Q] \tag{3.2}$$

is repeated in the respected paper by de Boer (1994/5): 'Every expression of a quantity Q consists of two factors or components. One of these is the name of a certain known *quantity* $[Q]$ of the same kind as the quantity to be expressed, which is taken as a standard of reference. The other component is the number of times $\{Q\}$ the standard is taken in order to make up the required quantity'.[3] We go one step further when pointing out that measurement units—as recognisable 'packets of measurement information'—play a role in communicating meaningful measurement information analogous to words in a language.

An interpretation of Eq. (3.2) is that any measurement of the quantity Q consists—as expressed by Maxwell 'making up'—of displacing the unit and counting how many times it fits in the measured displacement, where 'displacement' is not specifically in length, but in the dimension of interest. An assumption implicit in this procedure is that space is invariant in the dimension of the measured quantity, so that the unit as embodied in a measurement standard does not change on translation. The connexion with measurement units is made by the observation that $Q = \{Q\} \cdot [Q]$ relies on the invariance of the unit quantity upon measurement transformation. Conditions for invariance will be discussed in Sect. 3.2.3.

This interpretation of measurement units in terms of invariance is at a more philosophical and fundamental level than mere engineering. As mentioned above (Sect. 3.1.1), an engineer might see superficial similarities between Maxwell's Eq. (3.2) and for instance and expression for the volume, V, of combustion chamber above a combustion engine piston, $V = \varepsilon\ V_k$, where V_k – volume of combustion chamber when piston is in upmost position; ε – compression ratio (Kogan 2014). However, not all ratios of quantities reflect fundamental symmetries: Presumably in most applications there are more prosaic (albeit essential) factors, such thermal expansion, mechanical properties of piston material, etc., which affect the piston volume and need to be corrected for, and which are usually much larger than the effects of any fundamental symmetry breaking.

A case study of measurement units in the new SI (Sect. 3.6.1) will describe a connexion between measurement units and quantum mechanics, including unitary transformations. A second case study concluding this chapter (Sect. 3.6.2) will deal with measurement units in psychometry, with broader application in measurements in the social sciences.

[3]Note however that there is no explicit reference in Eq. (3.2) to which object/entity is being measured, but rather to a certain kind of quantity.

3.2.3 Symmetry, Conserved Quantities and Minimum Entropy. Maximum Entropy and Uncertainty

How much information that is carried by a word (or a unit, (Sect. 3.2.2)), or any other message, can be measured in terms of the concept of entropy in information theory (Shannon and Weaver 1963). Two distinct contexts where one seeks stationary (minimum or maximum) values of the entropy—that is, the amount of information—can be identified: (1) the best units for traceability are those with most order, i.e. least entropy, as an example of the principle of least action; (2) the change in entropy on transmission of measurement information cannot decrease, thus allowing realistic estimates of measurement uncertainty, in line with the second law of thermodynamics.

Firstly, symmetry—invariance under displacement—is a pre-requisite when forming easily recognised and 'meaningful' words and measurement units. The more symmetric—or ordered—a word is, the lower the entropy. Symmetry, having smaller entropy (most order), most meaning and unchanged on displacement, allows for a more efficient 'packaging' or 'chunking' of the information in a message, so the content is maintained on transmission and is more readily understood by a receiver. Attneave (1954), Barlow (2001), Miller (1956), Schneider et al. (1986) and others addressed how recognisable ('meaningful') patterns can improve communication by exploiting redundancy or 'chunking'. A basic concept in metrology is to choose a measurement unit which has the most order, i.e. least entropy, $H(P)$, associated with the message P. In terms of meaningfulness (Sect. 3.2.1), most information is contained in a message in which the entropy is minimised—which could be interpreted either in terms of the 'simplest' or most 'likely' signal. Meaningfulness is of course determined in each case by what the ultimate aim of the measurement is.

Figure 3.4 exemplifies a number of units of measure which are associated with recognisable patterns reflecting transformational symmetry (low entropy): a unit of time (**t**) for instance found in diverse physical systems (clock, atomic transition, planet, pulsar…, as in SI definition of second), where the canonical variable energy (**E**) is conserved under a 'displacement' through a 'distance'.

It is well known that the transformation symmetry sought when defining measurement units is related to conservation of the measured quantity. In seeking suitable systems with which to define and realise measurement units, one can observe that a number of physical quantities are known by experiment to be conserved in an isolated system: the total *energy, momentum, angular momentum* remain constant, whatever and however complex interactions occur within the system. These constancies are consequences of the invariance of mechanical systems under changes of corresponding canonical quantities—*time*, under *length* translation and under *rotation* in space, respectively—together with the principal of least action (Landau and Lifshitz (1976)). The role of entropy and symmetry in measurement is useful, not only in physics but also in the social sciences when seeking metrological standards for measurement. Examples include defining the difficulty of a cognitive

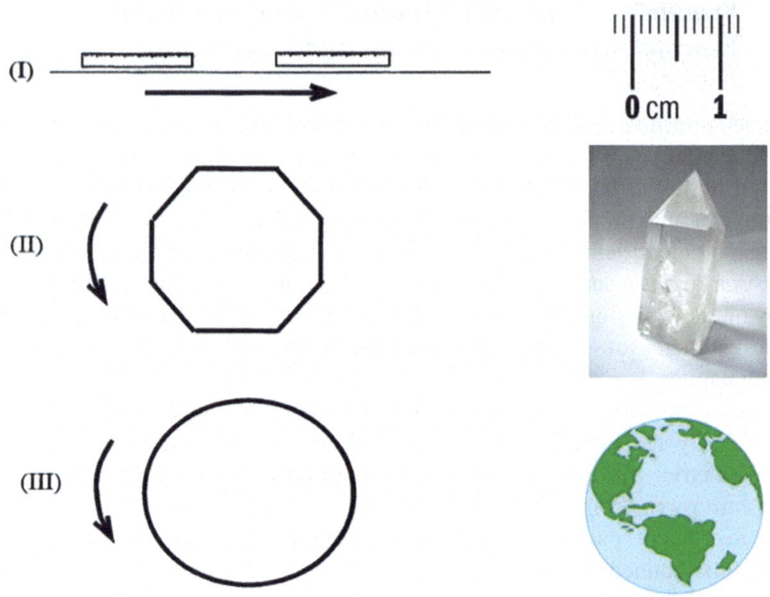

Fig. 3.4 Transformation symmetry and measurement units

task (in the Knox cube test (Sect. 5.3) or counting dots (Fig. 4.9)) where there is an obvious connexion: a more ordered task is easier to perform.

Secondly the principle of maximum entropy (most disorder) can be exploited when treating measurement error and uncertainty—so to say, the 'other side of the coin' to seeking the most ordered units. The loss of information on transmission through a measurement system can be modelled in terms of entropy (Eq. (3.3) in Sect. 5.4.1). Entropy can be in fact employed when characterising every element of a measurement system, for instance the commensurate ability of a person (or other probe) to perform the task. A more ordered person (less entropy) will be more able to perform, for instance the Knox block test[4] (Sect. 5.3).

In any process—for instance, transmission of measurement information through a measurement system—the change in entropy among all potential processes will be either zero or will increase as an example of the inexorable increase in disorder expressed by the second law of thermodynamics. Expressed mathematically:

$$H(Q|P) = H(P, Q) - H(P) \tag{3.3}$$

the entropy in the response, Q, of the measurement system given (conditional on) the stimulus input, P, is equal to the joint entropy of the message before and after

[4]This assumes of course that the necessary separation of object and instrument can be performed for the measurement system at hand (Sect. 3.1.2).

transmission minus the entropy of the original message. The joint entropy expresses how much information is transmitted and how much is lost or distorted (measurement uncertainty or bias) in the measurement process. The conditional entropy $H(Q|P)$ will be related (in Chap. 4, Eq. (4.5)) to a measure of the ability to perceive a dissimilarity in terms of a subjective distance between the distributions prior (P) and posterior (Q) to the measurement-based decision.

As explained by Klir and Folger (1988), the principle of maximum entropy applies when considering inferences whose content is beyond the evidence on hand. Otherwise called ampliative reasoning, one should 'use all but no more information than is available', that is, one should recognise and respect what one does not know—'knowing ignorance is strength' (Tsu 1972). Further discussion of measurement uncertainty will be found in Chap. 4 onwards.

3.3 Calibration, Accuracy and True Values

Measurement trueness goes beyond simply having small errors with respect to a local, perhaps arbitrary, reference value. Its essential meaning is in providing comparability of measurement results under both repeatability and reproducibility conditions—that is, measurements made by perhaps different laboratories using different equipment and operators. As society has developed and become more global, the need for measurement reference values of global applicability has increased in corresponding measure and a clear progression in metrological traceability to more universal measures is clear.

3.3.1 Trueness and Calibration Hierarchy

What defines a 'true' value, that is, the bull's eye of Fig. 2.2? Many years of discussion about the concept of measurement uncertainty (which is closely related but fundamentally different from metrological traceability (de Bièvre 2003)) has emphasised that—since true values are 'unknowable' and 'by nature indeterminate' quantities (to quote ISO GUM 1993 and VIM 2012, respectively)—then their use in measurement vocabularies is to be deprecated.

But in the context of metrological traceability there is one sense in which there is indeed access to what can be considered a true value—in fact in the very process of providing traceability, namely when making a calibration, that is, a measurement of a standard (or etalon) (Sect. 3.1.2). The value of a metrological standard (etalon), as determined previously with an uncertainty lower than in the actual measurements at hand, can be regarded as true—for instance, no one would consider the International Kilogram at the BIPM in Sèvres as giving anything but a true value of mass in any everyday weighing. The majority of measurements are made at lower levels in the

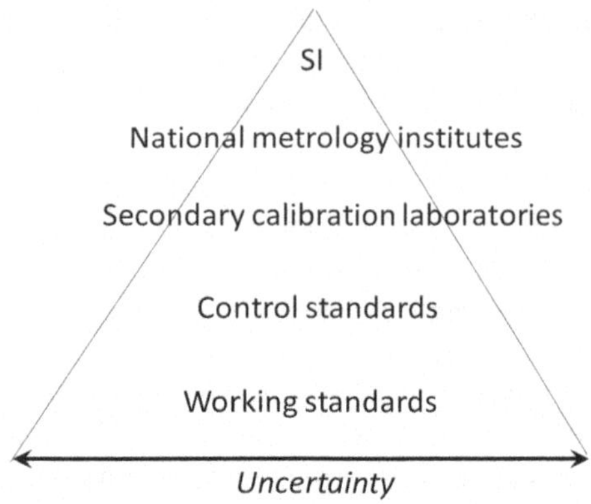

Fig. 3.5 Calibration hierarchy

calibration hierarchy[5] (Fig. 3.5) graded in order of measurement uncertainty and measures obtained from higher 'echelons' to all intents and purposes can be considered as true, since uncertainties in the standard are often much smaller than the uncertainty in the measurements at hand.

3.3.2 *Objectivity and Calibration of Instruments in the Social Sciences*

The process of calibration, i.e. tracing the measurement standards which, so to speak, embody defined amounts of the measurement unit, needs to be described in the case of a measurement system, including for the social sciences even systems where a human being is the instrument or other qualitative measurement situations.

In this book we identify two major factors which enable metrological traceability even in 'qualitative' situations such as found in the social sciences: (1) formulating measures of order in terms of entropy (Sect. 3.2.3) and (2) separating object and instrument (Sect. 3.1.2).

The Rasch model, mentioned in Sect. 1.4.5 and Eq. (1.3), is set up by performing a logistic regression to a set of data consisting of the responses of a number of instruments (e.g. people in a cohort) to a series of items (such as tasks).

A description of the calibration process where Man acts as measurement instrument in comparison with calibration of more traditional measurement systems in the physical sciences is given in Sect. 3.6.2.

While not enjoying access to universal units of measurement as in physics, in the social sciences—as in certified reference materials in analytical chemistry and materials science—one can establish recipes to define measurement units.

[5]Calibration hierarchy: 'sequence of calibrations from a reference to the final measuring system, where the outcome of each calibration depends on the outcome of the previous calibration' [VIM §2.40].

A construct specification equation (mentioned in the 'product' description Sect. 1.4.3 (Eq. (1.2))) will provide such a 'recipe' when formulating certified reference materials for traceability in the social sciences. A case study of the perception of the 'prestige' experienced with pre-packaged coffee will be evaluated in Sect. 4.5.2 as an example of the formulation of construct specification equations.

3.4 Politics and Philosophy of Metrology

3.4.1 Objective Measurement in the Physical and Engineering Sciences

The 'delicate' instruments of the physicist, referred to in *The Telegraphic Journal* 1884 (quoted in Gooday 1995), were not only used merely to make precise measurements (such as of the small electric currents in earlier telegraphy). The physicists' instruments also provided above all an 'absolute' accuracy, in other words a 'trueness', by which electrical quantities could be derived from the units of length, mass and time, the fundamental 'base' units of the metric system at that time. The universality and trueness of the latter were based on the ultimate physical reference of the era, namely the size and period of rotation of the Earth in true revolutionary universality '*A tous temps: A tous peuples*'. It took many years and was not until the electron was discovered at the turn of the nineteenth century before direct electrical measurements, with the voltmeter and ammeter, became to be raised in dignity and gain recognition as part of fundamental physics (Pendrill 2005, 2006a).

The same holds today in metrology: it is important to make precise measurements, in terms of low scatter or small uncertainty, as may be achieved by engineering a better measurement instrument. But perhaps arguably the main realm of the physicist in metrology is to provide for measurements which are traceable to absolute measures (ultimately, the universal fundamental constants). This enables the results of measurement to be related, not only of a particular quantity made by different people at different times and places (so important for trade and industry) but also to express different—apparently unrelated—quantities to each other in a more global sense. This latter universality of fundamental metrology relies on our understanding of the structure of the universe—spanning the realms of cosmology to elementary particle physics (Barrow 2002).

3.4.2 Politics and Trueness

In parallel with, and in some ways because of, the introduction of so-called neoliberal politics, views about how metrology was to be provided changed from about the 1980s, and this process is in a sense continuing for the foreseeable future.

A traditional view of a calibration hierarchy (Fig. 3.5) was where a national metrology institute (NMI) in each country (or perhaps the BIPM at the international level) would monopolise the top position at the pinnacle of the traceability pyramid for each measurement quantity. A more neoliberal approach would instead emphasise measurement laboratories of similar performance comparing their measurements, at a certain 'level' in the calibration hierarchy, such as a group of national metrology institutes aiming to refine the SI system for a particular measurement quantity. Previously exclusive consultative committees of the Metre Convention, had been the reserve of a few 'primary' NMIs charged with the tasking of recommending to the CIPM how each SI was to be defined. In the more neoliberal times, each consultative committee rapidly became open to many NMIs, albeit with some entry requirements demanding evidence of research.

'Customers', such as industrial metrology laboratories, would seek the 'best offer' from among a range of the NMIs as 'market actors'—a point of view which led by the turn of the Century to the Mutual Recognition Arrangement (MRA 1999) of the Metre Convention.

Another aspect of this neoliberal change in metrology was the increased emphasis on measurement uncertainty, to some extent at the expense of the more traditional fundamental concept of 'trueness'. Metrology was no longer to be 'only' furnishing unique true values, but instead one would allow for different, competing offers from whoever could convince the 'market'.

Over the years, NMIs have, as other organisations, been increasingly operated according to the neoliberal 'new public management (NPM) ' style, replacing former government 'authorities' which were perceived as inefficient and out-of-date. At the same, the private sector—which had pioneered this approach—increasingly found that NPM and neoliberalism were not optimal either...

3.4.3 Measurement Comparability in Conformity Assessment

Throughout the metrology vocabulary VIM, there is no mention of 'requirements' such as might be stipulated in conformity assessment, (apart from a few exceptions, such as 'maximum permissible measurement error' [VIM §4.26]). What might be called this 'general' perspective of measurement and quantities of the international metrology vocabulary is clear for instance in the Concept diagram for part of Clause 1 around 'quantity' of the VIM, where no explicit reference is made to either an entity or its quality characteristic. As mentioned in Sect. 2.2.2, one established area where conformity assessment of measurement instruments is regularly performed is legal metrology (Källgren et al. 2003; Pendrill 2014).

As explained in Chap. 1, the terminology of conformity assessment, in contrast to the metrological vocabulary, does emphasise a clear distinction (Fig. 1.3) between the quality characteristic $\eta = Z$ in the 'entity (or product) space' (Sect. 1.4.1) and the measurand of the quantity $\xi = Y$ in the 'measurement space' (Sect. 2.2.2). Metrological traceability provides the means of making measurement results comparable.

That measurement comparability is considered necessary to achieve the corresponding comparability of quality characteristics of entities, to the extent that such comparability is a requirement in conformity assessment. Comparability of entity quantities (or even kinds of quantity) is not always the prime concern in conformity assessment, where other factors—such as setting local safety limits—might be more pressing. In addition to the measurement value and any error in that estimate, it is also of interest in conformity assessment to deal with the variability—both real (in the entity) and apparent (as in measurement uncertainty). Uncertainty is not only expressed as a standard deviation, but in particular leads to risks of incorrect decisions of conformity. A pragmatic approach might be to set a limit on the measurement uncertainty, U_{cal}, associated with an uncorrected bias in relation to the tolerance maximum permissible error (*MPE*) in the context of testing for conformity assessment; perhaps stipulating that calibration uncertainties should not exceed half of the total measurement uncertainty, as an appropriate quantitative limit for when metrological traceability is significant in testing and measurement (Pendrill 2005, Sect. 4.3.2).

3.4.4 Objective Measurement in the Social Sciences

Physics has long been regarded as the original model for what a science should be. It has been a cherished hope and expectation of researchers in other disciplines that—given enough time, resources and talent—one should be able to achieve the same type of deep, broad and accurate knowledge achieved in the physical sciences, also in the biosciences or behavioural and social sciences. Research policy discussions often assume that all good science should be physics-like, i.e. characterized by the same quantitative specification of phenomena; a combination of mathematical sharpness and the deductive power of theory, and above all, a precise and profound understanding of the causes.

As Nelson (2015) has written: 'Whereas physics can limit the subject matter it addresses so that such heterogeneity is irrelevant to its aims, for other sciences, this diversity or variability is the essence of what they study.' As mentioned in Sect. 1.2.1, Nelson (2015) considers domains as diverse as cancer treatment, industrial innovation, K-12 education and environmental protection. Measurement uncertainty—as a measure of variability—is one the major hallmarks of metrology. We will later in this book give an account of a treatment of heterogeneity and the use of the concept of entropy when dealing with it.

Metrological references need to be founded on objective and sound measurement. Lacking an independent objective reality, e.g. in the social sciences, might lead to measurements providing no unique 'right' answer. This would make metrology of such qualitative assessments challenging, since: (1) independent reference standards used in metrology to ensure the comparability of different measurements would be difficult to establish separately from the actual measurement process, and

(2) measurement uncertainty would often be very large, since each new measurement set-up would produce definitions divergent from others.

In considering the philosophical foundations of social measurement, Maul et al. (2016) recall various approaches, including empiricism, pragmatism and realism. The philosophical realism behind physical metrology assumes, as in physics, that there is an objective reality, which exists even when we do not perceive or have instruments to measure it. One might argue—which Mari et al. (2016) refer to in terms of the output of their evaluation process—that there is 'seldom objective reality' in what is measured in social science (e.g. the challenge of a task) without our actually perceiving or measuring it. Mari et al. (2016) claim that a subjective opinion such as 'I am thirty percent happier today than I was yesterday' does not 'appear to deserve the trust that commonly accompanies measurement'.

Maul (2017), taking a broad perspective, argues that traditional approaches to the design and validation of survey-based measures may 'suffer from a number of serious shortcomings'. These include 'deeper confusions' regarding the relationship between psychological theory, modes of assessment and strategies for data analysis. Maul (2017) claims that operationalism may have encouraged the perception that psychological attributes need not be rigorously defined independently of a particular set of testing operations. He even states bluntly that the belief that measurement is a universally necessary component of scientific inquiry is 'simply not true'.

From our point of view, we would provide the counterargument that, for instance, opinions about fine art—e.g. the Mona Lisa painting by da Vinci—appear to be rather constant over the centuries and across different cultures. And measures of happiness are becoming essential components of person-centred care. A Rasch approach to perceived beauty or perceived happiness (or other pleasing patterns and degrees of order or symmetry) would in fact provide separate measures of the (albeit noisy) individual preferences of different persons and the intrinsic ability of, respectively, Leonardo's painting to stimulate pleasure or a particular activity of daily living to invoke happiness. This objectivity is perhaps not as strong as evaluations about the physical world (which would exist of course even without a human presence (Denbigh and Denbigh 1985)[6]), but is so to say 'fit for purpose' in the human-based context relevant for the present study. To use the vocabulary of the social sciences, such 'fit for purpose' references provide not only objectivity but—importantly—also intersubjectivity (Gillespie and Cornish 2010).

Our approach as presented in Sect. 3.3.2 can also be attributed to operationalism (as part of empiricism); i.e. defining a set of empirical operations performed with the measurement system. In that context, we circumvent the objections of realism since operationally it is 'meaningless to ask whether something is 'really' being measured' (Maul et al. 2016).

[6]'Definitional uncertainty'—'resulting from the finite amount of detail in the definition of a measurand' [VIM 2.27]—is of course in most cases much smaller in the strong objectivity of physics than in the social sciences.

In summary, the particular fusion of metrology and psychometrics proposed above, with its 'fit for purpose' objectivity and operationalism, appears to go some way in countering several of the philosophical reservations that had been expressed about attempting to quality-assure measurements in the social sciences (Pearce 2018).

3.5 Quantitative and Qualitative Scales

Table 3.3 summarises the different scales of Stevens, approximately compared with the data taxonomies suggested by Tukey. At the most elementary level, a categorical scale, called Nominal, is restricted to cases where no attempt is made to assign any order to the categories, which remain mere labels, being subordinate to all of the higher scales (ordinal, interval and ratio).

One step up from the nominal, towards a more quantitative scale is the case where, for intrinsic or extrinsic reasons, it is known that indications on a scale are generally ordered monotonically—i.e. a higher number indicates a higher measured quantity, although the exact, mathematical distance between marks on the scale is perhaps not known or investigated. Mathematically an ordinal scale can be expressed as:

$$x \geq y \; iff \varnothing(\mathrm{x}) \geq \varnothing(\mathrm{y})$$

Such, so-called 'ordinal' scales are subordinate to the more familiar interval and ratio scales; that is, even the latter have ordinal properties but enjoy more quantitative properties. Responses on an ordinal scale can be assigned to a series of discrete categories, as illustrated in Fig. 1.1, although such discrete scales are of course not unique to ordinality and even ordinal scales can be depicted as continuous, as in the so-called visual analogue scales.

Table 3.3 Scales and data taxonomies

Scales of measurement (Stevens 1946)	Data taxonomies (Mosteller and Tukey 1977, Chap. 5)
Ratio	*Balances* (unbounded, positive or negative values)
	Amounts (non-negative real numbers)
Interval	*Counts* (non-negative integers)
	Counted fractions (bounded by zero and one. Includes percentages, e.g.)
Ordinal	*Ranks* (starting from 1, which may represent either the largest or smallest)
	Grades (ordered labels such as Freshman, Sophomore, Junior, Senior)
Nominal	*Names*

The challenges of treating measurement responses on an ordinal scale have been known since at least the late nineteenth century but surprisingly, well over a hundred years later, it is still common to find users uncritically applying the usual tools of statistics—e.g. calculating means, standard deviations, confidence intervals, multivariate analysis of variance—to indicated values on an ordinal scale where of course those tools cannot be certain to be valid if the underlying scale distances are not exactly known.

Particularly relevant when treating categorical measurement in both the physical and social sciences are counted fraction scales (Sect. 3.5.1), as distinct from other ordinal scales (Sect. 3.5.2). Common examples of 'counted fractions' are performance metrics for ability tests, customer satisfaction and decision risks caused by uncertainty. Kaltoft et al. (2014) emphasised the importance of rating overall decision quality, for instance in patient-centred care. Examples include binary test methods, where the response of the measurement system is a simple 'yes' or 'no', as for example in chemistry and microbiology where the responses mean 'species identified' or not. Other qualitative measures include examination of images and patterns, e.g. in analytical chemistry (Hardcastle 1998; Ellison and Fearn 2005), forensics (Vosk 2015), healthcare (Mencattini and Mari 2015) and more generally the performance of diverse systems where a 'probe' is used to investigate an 'item', such as the efficiency (Fig. 4.11).

At a step further towards a more quantitative scale lie counts, that is non-negative integers as the most elementary example of an interval scale which will be mentioned in the first case study in Sect. 3.6.1.

3.5.1 Counted Fractions

One common family of ordinal scale is termed 'counted fractions', i.e. relative-number problems famously summarised by Tukey (1986), such as counting for example relative fractions of 'how many sheep & goats; are affected at this dose; or pebbles are quartz', and so on. As early as 1897, Pearson warned of the dangers of counting fractions: 'Beware of attempts to interpret correlations between ratios whose numerators and denominators contain common parts'. Mathematically, in the counted fraction expression, $X_j\% = \frac{X_j}{\sum_i X_i}$, the presence of the amount X of component j appearing in both the numerator and denominator means—and increasingly where X_j is either large or small compared with the other components—that any error in X_j will be correlated with the other components, since of course there is the boundary condition $\sum_j X_j\% = 100\%$. In the days before computers in the first half of the twentieth century, such non-linearities at the scale extremities—both the high and low ends—although known to be linearisable through so-called logistic ruling, converting percentages p into logits, l, with the expression $l = \ln\left(\frac{p}{100-p}\right)$, were

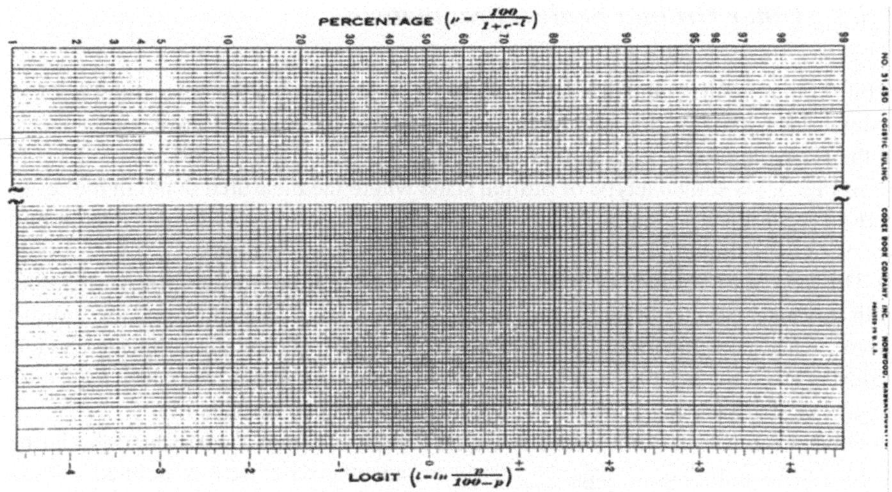

Fig. 3.6 Logistic ruling (Tukey 1986, reproduced with permission Taylor & Francis)

arduous to calculate, and it was common to use nomographs, such as exemplified in Fig. 3.6.

Throughout the century, the counted fraction dilemma has re-emerged in various schools. Aitchison (1982) for instance in what he termed compositional data analysis met with resistance in groups he termed:

Wishful Thinkers

No problem exists or, at worst, it is some esoteric mathematical statistical curiosity which has not worried our predecessors and so should not worry us. Let us continue to calculate and interpret correlations of raw components. After all if we omit one of the parts, the constant-sum constraint no longer applies. Someday, somehow, what we are doing will be shown by someone to have been correct all the time.

Describers

As long as we are just *describing* a compositional data set we can use *any* characteristics.
In describing compositional data we can use:

- arithmetic means,
- covariance matrices of raw components
- indeed any linear methods
- such as principal components of the raw components.

After all we are simply describing the data set in summary form, not analyzing it.

3.5.2 Other Ordinal Scales. Pragmatism

While common, the counted fraction scale (Sect. 3.5.1) is not the only example of an ordinal scale, and it is important to bear in mind that the logistic transformation used in the Rasch approach does not automatically fix all ordinality problems.

In Fig. 3.7 is shown a type of ordinal scale where the measured depth in metres of various geological strata do not of course directly correspond to the meaningful scales of deposition over different epochs and geological processes.

This and similar kinds of initially unknown scale may well be present together with the basic counted fraction ordinality, thus requiring special diagnostic tools to detect scale non-linearities and multidimensionality. Generalized linear models (GLM) (McCullagh 1980) can handle cases where the response variable (R) cannot always be expected to vary linearly with the explanatory variable (S).

Other examples are pragmatic scales (Table 3.1) where a cost function can be employed to introduce a distance metric on an otherwise weakly defined ordinal scale (Bashkansky et al. 2007), thus going beyond the traditional limitations of statistical measures of location and dispersion on such scales as well as of course capturing the impact of particular measurement 'values' in the broadest sense (Weinberger 2003). Examples may be found in legal metrology where commodities (such as petrol or pre-packaged goods) are normally priced linearly by quantity (Pendrill 2014). The impact of customer dissatisfaction on the other hand, for instance, with short measures of a commodity may well depend quadratically with increasing discrepancies, ε, of quantity from the expected value, as commonly expressed by the Taguchi loss function Cost $\cdot$ $(k \cdot \varepsilon)^2$, according to the impact of incorrect classification where 'Cost' is the cost per unit squared assignment error, ε^2, with respect to the expectation (or nominal value) (Pendrill 2006b). Such pragmatic

Fig. 3.7 Ordinal scale in geology. Photo: courtesy of Florence Pendrill

measures will be employed in the final chapter of this book when making decisions about product based on measurement.

3.6 New and Future Measurement Units

3.6.1 The Revised SI

There is much written elsewhere about the revised SI, for instance, as referenced on the Metre Convention website (BIPM 2018), to which we refer the reader for a more comprehensive account.

Classic examples of the universality of measurements in the framework of the laws of physics are the estimations of fundamental physical constants, such as the elementary electronic charge or the Boltzmann constant (Pendrill 1994). Completely different physical experiments which at first appearance are apparently unrelated—such as measurements of voltage in a semiconductor at cryogenic temperatures and a measurement made of the magnetic moment of a single electron in an electrostatic trap—can nevertheless yield consistent estimates of a particular fundamental physical constant. Traceability in the framework of the laws of physics thus enables apparently unrelated measurement quantities to be compared (Quinn 1994). The least-squares adjustment of the values of fundamental physical constants, as made periodically by the CODATA Task Force, is based on this (CODATA 2004).

De Boer (1994/5) writes: 'It is important to note that Maxwell writes that the *unit has also to be conceived as a quantity…*'. This statement does not need reassessment even though, as in the revised SI (CGPM 2018, 2019), the majority of SI units are defined in terms of fundamental constants. A careful choice of terminology would be, for example: 'the speed of light in vacuum *c has the value* is 299 792 458 m/s, …' (adapted from CGPM 2018).

Table 3.1 indicates how measurement units are expressed depending on which level in the hierarchy of concepts in quantity calculus is most relevant and meaningful to the task at hand.

Clear Definitions

A correct terminological formulation of measurement units in the SI system is particularly important, whose very role is to communicate measurement information meaningfully and widely; as summarised by Petley (1990) to everyone 'from the Nobel Prize winner to the proverbial man and woman in the street'. In the revised SI from 2018, a distinction is made between the new 'explicit constant'[7] and a more 'explicit unit' traditional definitions of units. A challenge posed by the explicit constant definitions of the revised SI is that they might be perceived as somewhat more abstract than the more traditional explicit unit definitions for some readers not familiar with fundamental physical constants.

Reasons for this can be that the revised definitions are formulated (1) in terms of the least informative concept level in quantity calculus, namely as 'quantity values' opposed to 'quantities' (Mari et al. 2018), and (2) assume a certain pre-requisite knowledge: In a table of concepts expected to be included in the terminology of a unit definition, several cells in the explicit constant definition table are left blank (–):

Explicit constant definition: unit of length

Unit	Kind of quantity	Quantity	Expression, equation of physics	Unit entity[a]	Values of constants	Related quantities	Related units
Metre (m)	Length	–	–	–	$c = 299\,792\,458\ m/s$	Time (t)	Second

[a]Maxwell's 'individual thing' (de Boer 1994/5)

These cells become filled in the more readily understandable explicit unit definition table:

Explicit unit definitions: Unit of length

Unit	Kind of quantity	Quantity[a]	Expression, equation of physics	Unit entity	Values of constants	Related quantities	Related units
Metre (m)	Length	Pathlength (ℓ)	$\ell = c \cdot t$	Path of light in vacuum	$c = 299\,792\,458\ m/s$	Time (t)	Second
Metre (m)	Length	Wavelength (λ)	$\lambda = \frac{c}{\nu}$	Wavelength of light in vacuum	$c = 299\,792\,458\ m/s$	Frequency (ν)	1/s

[a]According to quantity calculus, only quantities of the same kind have the same unit

In motivating the new approach of explicit constant definitions of the revised SI it is claimed that: 'A user is now free to choose any convenient equation of physics that links the defining constants to the quantity intended to be measured' (SI Brochure, 9th edition, 2.3.2). Two examples of such expressions are given in the explicit unit definition table above. The freedom of choice and the increased accuracy of fundamental physical constants offered by explicit constant definitions appears however to have to be traded against an increased difficulty—particularly for those unfamiliar with fundamental physics—in conveying meaning and understanding.

[7]'That is, a definition in which the unit is defined indirectly by specifying explicitly an exact value for a well-recognized fundamental constant' [24th CGPM, 2011 On the possible future revision of the International System of Units, the SI (CR, 532), Resolution 1].

Quantum Mechanics and Measurement

Dirac (1992) makes a brief mention of the special case when the real dynamical variable is a number, every state is an eigenstate and the dynamical variable is obviously an observable.

Any measurement of it always gives the same result, so it is just a physical constant, like the charge on an electron. A physical constant in quantum mechanics may thus be looked upon:

- either as an observable with a single eigenvalue
- or as a mere number appearing in the equations,

the two points of view, according to Dirac, being equivalent.

In line with our discussion of measurement units in the context of symmetry and entropy (Sect. 3.2.3), we would like to go beyond regarding a physical constant as a 'mere number'. One can attempt to describe the amount of information in terms of a sum of 'chunks' of the information in a message (Sect. 3.2.3), that is one seeks with a construct specification equation (Stenner et al. 1983, 2013; Sect. 1.4.3 (Eq. (1.2))) to resolve the information in the message into a number of explanatory variables which are the 'irreducible representations' corresponding to the eigenvectors, such as a series of spherical harmonics used to describe an arbitrary signal.

A connexion can be readily made between quantity calculus (Sect. 3.1) and the matrix formulation of quantum mechanics and the corresponding formulation in classical mechanics.[8]

Measurement is frequently mentioned when describing quantum mechanics, often referring to the impossibility—because of the non-zero value of the Planck constant—of making a measurement without disturbing the measurement object. This famously includes the Heisenberg uncertainty relation and contemporary topics in physics such as quantum entanglement and the possibilities of making quantum computers.

In the present context, we would like to highlight at the basic level how the measurement process is described in quantum mechanics, as a template for a description of measurement more broadly, in line with the aim of this book to give a unified view of measurement in the physical and social sciences. Two aspects in particular will be highlighted: (1) the distinct concepts of measurement object, quantity and quantity value and (2) measurement as a displacement process.

When discussing a general physical interpretation of a mathematical theory of quantum mechanics, Dirac (1992, §12, p. 45) recalls that in 'classical mechanics an observable always "has a value" for any particular state of the system' and clearly mentions measurement aspects: 'When we make an observation we measure some

[8]By the correspondence principle, relations specific to quantum mechanical effects, e.g. on the microscopic scale, can find correspondence to relations in Newtonian physics, e.g. at the macroscopic scale where the Planck constant is negligibly small.

dynamical variable. It is obvious physically that the result of such a measurement must always be a real number,...'

Eigenstates play of course a well-known role in quantum mechanics, and when introducing the expression:

$$\mathbf{Q}|q\rangle = q|q\rangle \tag{3.4}$$

Dirac (1992, §10, p. 35) emphasises the measurement aspects: 'If the dynamical system is in an eigenstate of a real, dynamical variable **Q**, belonging to the eigenvalue q, then a measurement of **Q** will certainly give as result the number q....'. Extension of this formulation to include functions of observables is described in §11 Dirac's (1992) book.

Dirac points out (1992, §12, p. 46) that one can 'extract physical information from the mathematics even when we are not dealing with eigenstates' provided one assumes that, 'if the measurement of the observable **Q** for the system in the state corresponding to $|q\rangle$ is made a large number of times, the average of all the results obtained will be $\langle q|\mathbf{Q}|q\rangle$, provided $|q\rangle$ is normalised.'

Interestingly, Dirac's formula (3.4) clearly includes explicit reference to the *entity* being measured by including $|q\rangle$. One can compare this with our description of measurement systems (Chap. 2). Note that the Dirac formulation also clarifies that **Q** on the LHS of (3.4) denotes a *quantity*—namely, the measurand associated with the measurement object (entity), as distinct from a *quantity value* (which would be the eigenvalue, q, on the RHS of (3.4)). This can be set in relation to our discussion of the hierarchy of measurement concepts in quantity calculus (Sect. 3.1.1 and Table 3.2).

A second aspect we would like to highlight is how *measurement as a displacement process* enters into a description of quantum mechanics. Recall Maxwell's classical words in connexion with Eq. (3.2) of 'making up' a required quantity, and counting how many times a unit fits in the measured displacement, where 'displacement' is not specifically in length, but in the dimension of interest. The aim of the displacement can be interpreted in the broadest sense, not just in the physicist's laboratory, but more generally including the measurement comparability needed in any application (Sect. 3.2).

Dirac (1992, §25) writes about displacement operators which can be seen as a description of a measurement process: 'The displacement of a state or observable is a perfectly definite process physically. Thus to displace a state or observable through a distance δx in the direction of the x-axis, we should merely have to displace all the apparatus used in preparing the state, or all the apparatus required to measure the observable, through the distance δx in the direction of the x-axis, and the displaced apparatus would define the displaced state or observable. ... A displaced state or dynamical variable is uniquely determined by the undisplaced state or dynamical variable together with the direction and magnitude of the displacement'.

Dirac (1992, §26) goes on to describe a displacement operator, D, such that for any dynamical variable v, a displaced dynamical variable $v_d = DvD^{-1}$. One particular kind of displacement operator is a linear *unitary* transformation operator U. This special displacement operator can transform any linear operator Q to a corresponding linear operator $Q*$ according to the relation: $Q^* = U \cdot Q \cdot U^{-1}$ so that each $Q*$ has the *same* eigenvalues as the corresponding Q, $Q|q\rangle = q|q\rangle$ and hence: $Q^* U|q\rangle = qU|q\rangle$.

A connexion between Dirac's and Maxwell's descriptions of displaced measurement systems can be established by observing that the unit quantity has an eigenvalue q_{unit}: $[Q]|q\rangle = \{q_{\text{unit}}\}|q\rangle$, and a 'displaced' observable: $Q^*|q\rangle = \{Q\} \cdot [Q]|q\rangle = \{Q\} \cdot \{q_{\text{unit}}\}|q\rangle$, by simply combining (3.4) and (3.2), so that:

$$U = \{q_{\text{unit}}\} \cdot [Q] \tag{3.5}$$

The quantisation rule: $\oint p \cdot dq = n \cdot h$ for a pair of canonical variables (p, q) such as position and momentum or time and energy (Born 1972), means that, in one period of the motion, the integral yields an area which is an integral multiple of h, according to the quantum postulate. The eigenfunctions of orbital angular momentum operator, as irreducible representations, form a set of base functions accompanied by a set of eigenvalues (multiples of h) which can be regarded as measurement units (or 'chunks') (Sect. 3.2.3). For our purposes in metrology, it illustrates how the Planck constant h acts as a fundamental measurement unit.[9]

How the fundamental physical constants now enter into definitions of measurement units in the revised SI is described further in 9th ed SI Brochure (CGPM 2019, CGPM 2018, 2019). In the present description, one can seek units of measurement more generally from fundamental symmetries described in terms of minimum entropy (Sect. 3.2.3).

The examples given in the remaining sections of this chapter in different ways have to do with counting.

Boltzmann Constant and Elementary Counting

Atomic helium and its microscopic structure have been the focus of attention of a number of highly accurate experimental and theoretical investigations of properties such as electronic binding energies, fine structure, etc., which are leading in some cases to new estimates of fundamental constants, such as α, the fine structure constant. At the same time, the thermodynamic properties of *macroscopic* helium gas are being studied intensively, particularly in connection with the development of gas thermometry. Connecting these microscopic and macroscopic studies of helium,

---

[9]The Planck constant is not merely a 'number' but has multiplicity of roles, such as (1) a constant of proportionality between canonical pairs of quantities (e.g. energy/time, momentum/position, angular momentum/rotation); (2) acting as a fundamental unit as the 'quantum' of 'action' (e.g. energy·time); (3) is implicit in many of the 'quantum' definitions of the SI, not only the kilogram but also the second, volt and ohm; (4) quantifying the interaction through fields between physical systems, such as the electromagnetic interaction mediated by 'virtual' photons (Cohen-Tanoudji 1993).

Pendrill (1996) proposed a new estimate of the Boltzmann constant, k, as well as a general assessment of the reliability of optical and electrical measurements of the polarisation properties of helium gas. The link between the macroscopic (gas constant, R, from dielectric constant gas thermometry (DCGT) of ^{4}He) and the microscopic (Boltzmann constant) is provided by the relation: $R = N_A \cdot k$, where the Avogadro constant, N_A, represents counting the number of elementary microscopic entities to make up a macroscopic quantity. The 1996 estimate proposed that the Boltzmann constant, k had the value 1.380 628(16) $\times$ 10^{-23} J/K (provided one accepted the *ab initio* relativistic value of the atomic polarizability and the DCGT experimental results available at the time) can be compared with the most recent value proposed in the new SI: 1.380 649 $\times$ 10^{-23} J/K (CGPM 2018, 2019).

Counts and Quantities of Unit One

An example of elementary counting (of the number of dots) regarded as measurement is given in Fig. 4.9.

Whether 'counts' (non-negative integers), of for example the number of pills, can be considered as 'dimensionless quantities in the SI' is still an active subject of debate in the international literature (Flater 2017). As remarked by Kogan (2014), a 'quantity for which all exponents of factors corresponding to base quantities in its quantity dimension are zero', is preferably called a quantity of unit one (where 'dimension' is discussed in Sect. 3.2.2).

The introduction of the number of entities as a base quantity can help answer the question: 'is amount of information a quantity?' (Kogan 2014) The amount of information usually stands for measure of information in a message. This quantity has its own unit in the informatics—it is called bit and defined as a unit of the amount of information in the binary computing, the minimal unit of the amount of information that can be transmitted or stored, and corresponds to one binary digit that can have one of the two values, 0 or 1. One of the earliest and elementary measures of information was formulated by Hartley (Klir and Folger 1988) in terms of how specific a particular sign is:

$$I(N) = \log_b(N) \tag{3.6}$$

where N is the number of distinguishable signs. The least amount (a 'quantum') of information—one bit—corresponds to $N = 2$ signs (e.g. 0 & 1) on a base of $b = 2$. Equation (3.6) can be converted into terms of informational entropy by multiplying by the Boltzmann constant, k, which, together with the bit, play a role in 'information quantisation' analogous to that of the Planck constant in quantum mechanics (Cohen-Tannoudji 1993). We return to the connexion between entropy and information in our studies of cognitive tests in neuropsychology and elementary counting (Chap. 4).

3.6.2 *Human Challenges*

In its logistic regression form, the 'straight ruler' aspect of the Rasch formula, i.e. Eq. (1.1), has been described by Linacre and Wright (1989) in the following terms: 'The mathematical unit of Rasch measurement, the log-odds unit or "logit", is defined prior to the experiment. All logits are the same length with respect to this change in the odds of observing the indicative event.'

The Rasch invariant measure approach goes further in defining measurement units (Humphry 2011) since it uniquely yields estimates 'not affected by the abilities or attitudes of the particular persons measured, or by the difficulties of the particular survey or test items used to measure', i.e. specific objectivity (Irwin 2007). The Rasch (1961) approach is not simply mathematical or statistical, but instead a specifically metrological approach to human-based measurement. Note that the same probability of success can be obtained with an able person performing a difficult task as with a less able person tackling an easier task. The separation of attributes of the measured item from those of the person measuring them brings invariant measurement theory to psychometrics. Fisher's (1997) work on the metrology of instruments for physical disability (Fig. 3.8) was one of the first to demonstrate the concepts of intercomparability through common units, where item banking is a common expression (Pesudovs 2010).

Having enabled with Rasch Measurement Theory a set of metrological references, e.g. for task difficulty, one can then proceed to set up a scale (analogous to conventional measurement etalons (Fig. 3.9)) which is delineated by measurement units where any measured quantity, $\delta_j = \{\delta_j\} \cdot [\delta]$, is the product of a number { } and a unit denoted in square brackets [ ]—Eq. (3.2). This step is enabled by combining a

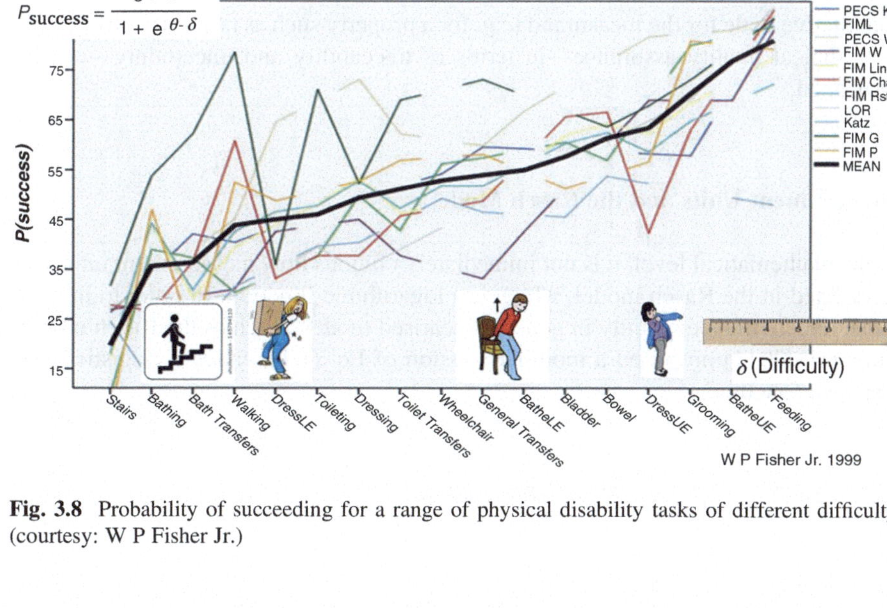

Fig. 3.8 Probability of succeeding for a range of physical disability tasks of different difficulty (courtesy: W P Fisher Jr.)

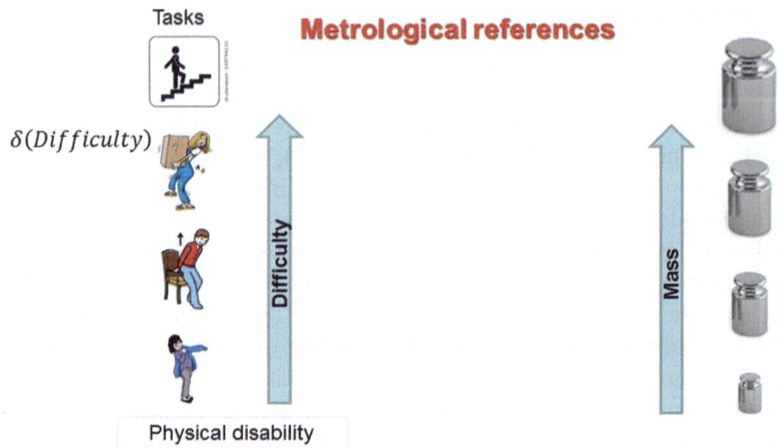

Fig. 3.9 A set of tasks of increasing difficulty as metrological references analogous to a set of mass standards

procedure to transform qualitative data to a new 'space' (in the present case, through restitution, to the space of the measurand, as illustrated in Figs. 2.11 and 3.3), together with ability of Rasch Measurement Theory to provide separate estimates of measurement and object dispersions in the results when Man acts as a measurement instrument.

This new approach to the metrological treatment of qualitative data differs from others in that the special character of the qualitative data is assigned principally not to the measurand but to the response of the measurement system. (One can draw analogies with the common expression: 'Beauty is in the eye of the beholder'.) Using Rasch Measurement Theory in the restitution process re-establishes a linear, quantitative scale for the measurand (e.g. for a property such as task difficulty) where metrological quality assurance—in terms of traceability and uncertainty—can be performed.

Measurement Units and the Rasch Model

At the mathematical level, it is not immediately obvious how measurement units can be inserted in the Rasch model, which is a logarithmic function. In order to include measurement units explicitly in general linearised models, such as the Rasch model, Humphry (2011) proposed a modified version of Eq. (1.1), called a 'logistic measurement function':

$$\ln\left(\frac{P_{\text{success},\,i,j}}{1 - P_{\text{success},\,i,j}}\right) = \rho_s \cdot \left(\theta_i^* - \delta_j^*\right) \tag{3.7}$$

where s indicates a classification of an empirical factor; ρ is a multiplicative constant; and the modified Rasch parameters are related to the original Rasch parameters through the expressions $\theta_i^* = \frac{\theta_i}{\rho}$ and $\delta_j^* = \frac{\delta_j}{\rho}$. If units are to be associated with person and item attributes, respectively, as $\theta_i = \{\theta_i\} \cdot [\theta]$ and $\delta_j = \{\delta_j\} \cdot [\delta]$ then assuming that item and person attributes share the same scale—a key aspect of the Rasch model—gives an expression for the 'common unit' of measure as: $[\theta] = [\delta]$ (denoted $[u_*]$ by Humphry (2011)).

Equation (3.7) appears on first sight to be similar to Item Response Theory expressions, but there is a subtle distinction, as expressed recently: 'Item Response Theory models are statistical models used to explain data, and the aim of an Item Response Theory analysis is to find the statistical model that best explains the observed data. By contrast, the aim of Rasch Measurement Theory is to determine the extent to which observed clinical outcome assessment data satisfy the measurement model' (Barbic and Cano 2016).

As pointed out by Humphry and Andrich (2008), the incorporation in an Item Response Theory model of a discrimination parameter which is estimated for each item (or person) will in general break conditions for sufficiency and specific objectivity, and thus the opportunity of establishing units and measurement scales. But this opportunity is maintained if one, as in Eq. (3.7), associates a discrimination factor (ρ) with a *set* of items rather than a single item, according to Humphry and co-workers (Humphry 2011; Asril and Marais 2011), as will be exemplified in Sect. 5.4.1.

The concept of entropy will be a main guide in this book when formulating construct specification equations, be it of task difficulty or person (instrument) ability (Sect. 5.1). The increasing difficulty of remembering block sequences (such as the Corsi block test) was described by Schnore and Partington (1967) in terms of a sum of a set of basic patterns (or chunks) with different information content expressed in terms of entropy and the symmetry of each pattern. These can be regarded as measurement units in a similar fashion to the irreducible representations used to describe arbitrary functions such as in quantum mechanics (Sects. 3.2.3 and 3.6.1).

References

J. Aitchison, The statistical analysis of compositional data. J. R. Stat. Soc. **44**, 139–177 (1982)

A. Asril and I. Marais, (2011) Applying a Rasch Model Distractor Analysis. In: Cavanagh R.F., Waugh R.F. (eds) Applications of Rasch Measurement in Learning Environments Research. Advances in Learning Environments Research, vol **2**. SensePublishers, Rotterdam

S. Barbic and S. Cano, The application of Rasch measurement theory to psychiatric clinical outcomes research, BJPsych Bulletin, **40**, 243–244, https://doi.org/10.1192/pb.bp.115.052290 (2016)

F. Attneave, Informational aspects of visual perception. Psychol. Rev. **61**, 183–193 (1954)

H. Barlow, The exploitation of regularities in the environment by the brain. Behav. Brain Sci. **24**, 602–607 (2001)

J. Barrow, *From Alpha to Omega* (Jonathan Cape, London, 2002). ISBN 0224061356

E. Bashkansky, S. Dror, R. Ravid, P. Grabov, Effectiveness of a product quality classifier. Qual. Eng. **19**(3), 235–244 (2007)

J.P. Bentley, *Principles of Measurement Systems*, 4th edn. (Pearson, Prentice-Hall, Lebanon, 2004). ISBN-13: 978-0130430281, ISBN-10: 0130430285

BIPM, On the Future Revision of the SI, (2018), https://www.bipm.org/en/measurement-units/rev-si/

M. Born, *Atomic Physics*, 8th edn. (Blackie & Son Ltd., London, 1972). 216.89027.6, ISBN-13: 978-0486659848, ISBN-10: 0486659844

N.R. Campbell, *Physics – The Elements* (Cambridge University Press, Cambridge, 1920)

CGPM, SI Brochure: The International System of Units (SI), 9th edn. (2019), https://www.bipm.org/utils/common/pdf/si-brochure/SI-Brochure-9.pdf

CGPM, On the revision of the International System of Units (SI), in *Draft Resolution* **A** – 26th meeting of the CGPM (13–16 November 2018), (2018), https://www.bipm.org/utils/en/pdf/CGPM/Draft-Resolution-A-EN.pdf

CODATA, *Task Force on Fundamental Physical Constants, Committee on Data for Science and Technology* (ICSU, Paris, 2004). http://www.codata.org/taskgroups/TGfundconst/index.html

G. Cohen-Tannoudji, Universal Constants in Physics (MCGRAW HILL HORIZONS OF SCIENCE SERIES), ISBN-13: 978-0070116511, McGraw-Hill Ryerson, Limited (1993)

P. de Bièvre, Traceability is not meant to reduce uncertainty. Accred. Qual. Assur. **8**, 497 (2003)

J. de Boer, On the history of quantity calculus and the international system. Metrologia **31**, 405–429 (1994/5)

K.G. Denbigh, J.S. Denbigh, *Entropy in Relation to Incomplete Knowledge* (Cambridge University Press, Cambridge, 1985). ISBN 0 521 25677 1

P.A.M. Dirac, The principles of quantum mechanics, in *The International Series of Monographs on Physics*, ed. by J. Birman et al., 4th edn., (Clarendon Press, Oxford, 1992)

S. Ellison, T. Fearn, Characterising the performance of qualitative analytical methods: statistics and terminology. TRAC-Trend Anal. Chem. **24**, 468–476 (2005)

W.H. Emerson, On quantity calculus and units of measurement. Metrologia **45**, 134–138 (2008)

EN 15224:2012, *Health care services – Quality management systems – Requirements based on EN ISO 9001:2008*

W.P. Fisher, Jr., Physical disability construct convergence across instruments: Towards a universal metric. Journal of Outcome Measurement **1**(2), pp 87–113 (1997)

R. Feynman, The Feynman Lectures on Physics, vol. I, (2013), http://www.feynmanlectures.caltech.edu/I_01.html#Ch1-S1

D. Flater, Redressing grievances with the treatment of dimensionless quantities in SI. Measurement **109**, 105–110 (2017)

R. Fleischmann, Einheiteninvariante Gröβengleichungen, Dimension. Der Mathematische und Naturwissenschaftliche Unterricht **12**, 386–399 (1960)

A. Gillespie, F. Cornish, Intersubjectivity: towards a dialogical analysis. J. Theory Soc. Behav. **40**, 19–46 (2010)

G. Gooday, in *The Values of Precision*, ed. by M. N. Wise, (Princeton University Press, Princeton, 1995). ISBN 0-691-03759-0

W. Hardcastle, *Qualitative Analysis: A Guide to Best Practice* (Royal Society of Chemistry, Cambridge, 1998)

S.M. Humphry, The role of the unit in physics and psychometrics. Meas. Interdiscip. Res. Perspect. **9**(1), 1–24 (2011)

S.M. Humphry, D. Andrich, Understanding the unit implicit in the Rasch model. J. Appl. Meas. **9**, 249–264 (2008)

R.J. Irwin, A psychophysical interpretation of Rasch's psychometric principle of specific objectivity. Proc. Fechner Day **23**, 1–6 (2007)

JCGM200:2012 International vocabulary of metrology—basic and general concepts and associated terms (VIM 3rd edition) (JCGM 200:2008 with minor corrections) WG2 Joint Committee on Guides in Metrology (JCGM) (Sevrès: BIPM)

H. Källgren, M. Lauwaars, B. Magnusson, L.R. Pendrill, P. Taylor, Role of measurement uncertainty in conformity assessment in legal metrology and trade. Accred. Qual. Assur. **8**, 541–547 (2003)

M. Kaltoft, M. Cunich, G. Salkeld, J. Dowie, Assessing decision quality in patient-centred care requires a preference-sensitive measure. J. Health Serv. Res. Policy **19**, 110–117 (2014). https://doi.org/10.1177/1355819613511076

G.J. Klir, T.A. Folger, *Fuzzy sets, uncertainty and information* (Prentice Hall, New Jersey, 1988). ISBN 0-13-345984-5

J. Kogan, An Alternative Path to a New SI, Part 1: On Quantities with Dimension One, (2014), https://web.archive.org/web/20160912224601/http://metrologybytes.net/PapersUnpub/Kogan_2014.pdf

L.D. Landau, E.M. Lifshitz, Mechanics, Course of Theoretical Physics (Butterworth-Heinenann, 1976), 3rd ed., Vol. 1. ISBN 0-7506-2896-0.

B. Langefors, Essays on Infology, in *Gothenburg Studies of Information Systems*, ed. by B. Dahlbom. Report **5** (University of Göteborg, 1993)

J.M. Linacre, B. Wright, The 'length' of a Logit. Rasch Meas. Trans. **3**, 54–55 (1989)

L. Marco-Ruiz, A. Budrionis, K.Y. Yigzaw, J.G. Bellika, Interoperability Mechanisms of Clinical Decision Support Systems: A Systematic Review, in *Proceedings of the 14th Scandinavian Conference on Health Informatics*, Gothenburg, Sweden, April 6–7, 2016, (2016). http://www.ep.liu.se/ecp/122/ecp16122.pdf

L. Mari, A. Maul, D. Torres Irribarra, M. Wilson, A metastructural understanding of measurement. J. Phys. Conf. Ser **772**, 012009 (2016). IMEKO2016 TC1-TC7-TC13

L. Mari, C.D. Ehrlich, L.R. Pendrill, Measurement units as quantities of objects or values of quantities: a discussion. Metrologia **55**, 716 (2018). https://doi.org/10.1088/1681-7575/aad8d8

A. Maul, D. Torres Irribarra, M. Wilson, On the philosophical foundations of psychological measurement. Measurement **79**, 311–320 (2016)

A. Maul. Rethinking traditional methods of survey validation, Measurement. Interdisciplinary Research and Perspectives **15**(2), 51–56 (2017)

P McCullagh, Regression models for ordinal data. J. Roy. Stat. Soc., 42: p. 109–42 (1980)

A. Mencattini, L. Mari, A conceptual framework for concept definition in measurement: the case of 'sensitivity. Measurement **72**, 77–87 (2015)

G.A. Miller, The magical number seven, plus or minus two: some limits on our capacity for processing information. Psychol. Rev. **63**, 81–97 (1956)

F. Mosteller, J.W. Tukey, *Data Analysis and Regression: A Second Course in Statistics* (Addison-Wesley, Reading, 1977)

MRA, International equivalence of measurements: the CIPM MRA, (1999), https://www.bipm.org/en/cipm-mra/

R.R. Nelson, Physics envy: get over it. Iss. Sci. Technol. **XXXI**, 71–78 (2015). http://issues.org/31-3/physics-envy-get-over-it/

J. Pearce, Psychometrics in action, science as practice. Adv. Health Sci. Educ. **23**, 653–663 (2018). https://doi.org/10.1007/s10459-017-9789-7

L.R. Pendrill, Assuring measurement quality in person-centred healthcare. Meas. Sci. Technol **29**(3), 034003 (2018). https://doi.org/10.1088/1361-6501/aa9cd2. special issue Metrologie 2017.

L.R. Pendrill, Some comments on fundamental constants and units of measurement related to precise measurements with trapped charged particles. Phys. Scripta **T59**, 46–52 (1994). (1995) and World Scientific Publishing, ed. I Bergström, C Carlberg & R Schuch, ISBN 981-02-2481-8 (1996)

L.R. Pendrill, Macroscopic and microscopic polarisabilities of helium gas. J. Phys. B At. Mol. Opt. Phys. **29**, 3581–3586 (1996)

L.R. Pendrill, Meeting future needs for metrological traceability – a physicist's view. Accred. Qual. Assur. **10**, 133–139 (2005). http://www.springerlink.com/content/0dn6x90cmr8hq3v4/?p=2338bc01ade44a208a2d8fb148ecd37a&pi

L.R. Pendrill, Metrology: time for a new look at the physics of traceable measurement? Europhysics News **37**, 22–25 (2006a). https://doi.org/10.1051/epn:2006104

L.R. Pendrill, Optimised measurement uncertainty and decision-making when sampling by variables or by attribute. Measurement **39**(9), 829–840 (2006b). https://doi.org/10.1016/j.measurement.2006.04.014

L.R. Pendrill, Uncertainty & risks in decision-making in qualitative measurement, in *AMCTM 2011 International Conference on Advanced Mathematical and Computational Tools in Metrology and Testing*, Göteborg June 20–22 2011, (2011), http://www.sp.se/AMCTM2011

L.R. Pendrill, Using measurement uncertainty in decision-making & conformity assessment. Metrologia **51**, S206 (2014)

K. Pesudovs, Item banking: a generational change in patient-reported outcome measurement. Optom. Vis. Sci. **87**(4), 285–293 (2010). https://doi.org/10.1097/OPX.0b013e3181d408d7

B.W. Petley, *The fundamental physical constants and the frontier of measurement* (Adam Hilger Ltd, Bristol, 1985). ISBN 0-85274-427-7

B.W. Petley, Thirty years (or so) of the SI. Meas. Sci. Technol. **1**, 1261 (1990). https://doi.org/10.1088/0957-0233/1/11/023

T. J. Quinn, Metrology, its role in today's world BIPM Rapport BIPM-94/5 (1994)

A.P. Raposo, The Algebraic Structure of Quantity Calculus, (2016), arXiv:1611.01502v1

G. Rasch, On general laws and the meaning of measurement in psychology, in *Proceedings of the Fourth Berkeley Symposium on Mathematical Statistics and Probability*, (University of California Press, Berkeley, 1961), pp. 321–334

F.S. Roberts, Measurement theory with applications to decision-making, utility, and the social sciences, in *Encyclopedia of Mathematics and Its Applications*, vol. 7, (Cambridge University Press, Cambridge, 1985). ISBN 978-0-521-30227-2

G.B. Rossi, Measurement and probability – a probabilistic theory of measurement with applications, in *Springer Series in Measurement Science and Technology*, (2014), https://doi.org/10.1007/978-94-017-8825-0

T.D. Schneider, G.D. Stormo, L. Gold, A. Ehrenfeuch, The information content of binding sites on nucleotide sequences. J. Mol. Biol. **188**, 415–431 (1986). www.lecb.ncifcrf.gov/~toms/paper/schneider1986

M.M. Schnore, J.T. Partington, Immediate memory for visual patterns: symmetry and amount of information. Psychon. Sci. **8**, 421–422 (1967)

C.E. Shannon, W.W.Weaver The mathematical theory of communications. University of Illinois Press, Urbana, 117 p. (1963)

K.D. Sommer, B.R.L. Siebert, Systematic approach to the modelling of measurements for uncertainty evaluation. Metrologia **43**, S200–S210 (2006). https://doi.org/10.1088/0026.1394/43/4/S06

A.J. Stenner, M. Smith III, D.S. Burdick, Toward a theory of construct definition. J. Educ. Meas. **20** (4), 305–316 (1983)

A.J. Stenner, W.P. Fisher Jr., M.H. Stone, D.S. Burdick, Causal Rasch models. Front. Psychol. **4** (536), 1–14 (2013)

S.S. Stevens, On the theory of scales of measurement. Science **103**(2684), 677–680 (1946). New Series

L. Tsu, *Tao Te Ching, Chapter 71* (Vintage Books (Random House), New York, 1972)

J.A. Tukey, Chapter 8, Data analysis and behavioural science, in *The Collected Works of John A Tukey, Volume III, Philosophy and Principles of Data Analysis: 1949 – 1964*, ed. by L. V. Jones, (University North Carolina, Chapel Hill, 1986)

P.F. Velleman, L. Wilkinson, Nominal, ordinal, interval, and ratio typologies are misleading. Am. Stat. **47**, 65–72 (1993). https://www.cs.uic.edu/~wilkinson/Publications/stevens.pdf

T. Vosk, Measurement uncertainty: requirement for admission of forensic science evidence, in *Wiley Encyclopedia of Forensic Science*, ed. by A. Jamieson, A. A. Moenssens, (Wiley, Chichester, 2015)

J. Wallot, Dimensionen, Einheiten, Massysteme, in *Handbuch der Physik II, Kap. I*, (Springer, Berlin, 1926)

W. Weaver, C. Shannon, *The Mathematical Theory of Communication* (University of Illinois Press, Champaign, 1963). ISBN 0252725484

E.D. Weinberger, A theory of pragmatic information and its application to the quasi-species model of biological evolution. Biosystems **66**, 105–119 (2003). http://arxiv.org/abs/nlin.AO/0105030

Chapter 4
Measurement

4.1 Performing Measurements

Implementation of a measurement method or measurement system can be regarded as being situated—at the point of 'measurement'—about halfway round the quality loop shown in Fig. 2.1. One needs, when assuring quality of product, to:

> . . . plan and implement those processes for monitoring, measurement, analysis . . . .which are needed to demonstrate that the product conforms to requirements. . . (ISO 9001 Management system for Quality—Requirements §8).

Characteristics to be checked when implementing a measurement method or system have been introduced in Sects. 2.5 and 2.6. Whether one aims at:

- one-off use of a method,
- the establishment and maintenance of a measurement process based on the method.

it is recommended to perform calibration and metrological confirmation prior to embarking on more extensive series of measurements in 'production'. The confirmation process will be described in Sect. 4.2.

The evaluation of measurement uncertainty is a key step, both in the metrological confirmation process as well as in subsequent measurements and decision-making, and will be reviewed in Sects. 4.2.2 and 4.4 for physical and social measurements, respectively.

How the concepts of calibration and traceability (introduced in Chap. 3) are regarded when performing measurement in the different disciplines, such as physics, engineering, chemistry and the social sciences, will be reviewed in Sect. 4.3. Section 4.4 will look in depth at metrological concepts in the social sciences.

Examples of the results of actually performing measurement spanning the physical and social sciences will round off this chapter (Sect. 4.5) to illustrate treatment of the results of implementing a measurement method or system, including a continuation of the example of pre-packaged goods chosen in this book. As before,

L. Pendrill, *Quality Assured Measurement*, Springer Series in Measurement Science and Technology, https://doi.org/10.1007/978-3-030-28695-8_4

templates are provided for the reader to complete the corresponding sections of the measurement task for their chosen case.

4.1.1 Measurement Process

Measurement—the 'process of experimentally obtaining one or more quantity values that can reasonably be attributed to a quantity' (VIM §2.1)—produces in the simplest case typically an estimate of a single measurand at a certain level (say a mass of 500 g or a product of certain quality) from the observed response of a measurement system previously calibrated against one measurement standard (etalon).

Measurements in the physical and social sciences are however seldom made only at one level with one standard. A more general and useful picture of a measurement process is shown in Fig. 4.1.

Three steps (Fig. 4.1) can be identified: (A) calibration of a set of standards; (B) calibration and testing of a measurement system at the various levels; (C) implementation to measure the value of the measurand at each level of interest. These three steps are taken when determining a physical quantity like mass at different levels such as 1 kg, 500 g and 100 g, or a more perceptual property characterising product quality such as smoothness, at different levels from rough to smooth; or task difficulty, from easy to difficult. Metrological performance will in

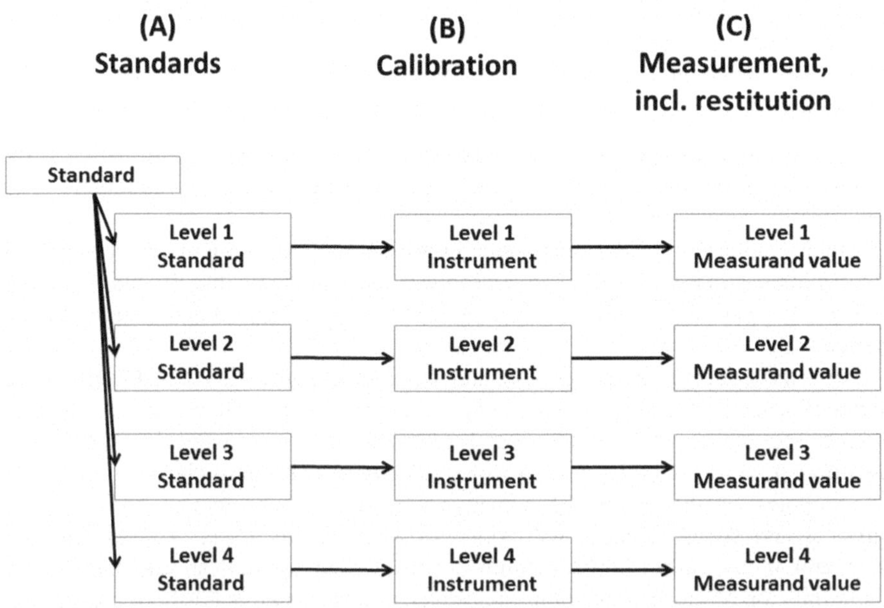

Fig. 4.1 Three steps when implementing a measurement method or system

general differ from level to level. The starting point or 'anchor' for the whole set of measurement, as shown on the far left of Fig. 4.1, will be access to a 'head-weight', that is, a principal measurement standard (etalon) where measurement accuracy is highest, as the result of a previous calibration higher up the traceability hierarchy (Sect. 3.3.1). Two case studies illustrate this general set-up in Sect. 4.5.

4.2 Metrological Confirmation

Confidence in the measurements performed in the conformity assessment of any entity (product, process, system, person, body or phenomenon) can be considered sufficiently important that the measurements themselves will be subject to the steps (a) to (f) of the product quality loop, that is, a 'measurement quality loop' where 'product' is a measurement result (Fig. 2.1). This is, so to say, a kind of metrological conformity assessment 'embedded' in any product conformity assessment, encompassing *Specification of demands on a measurement system* (Sect. 2.2), and *Decision-making* (Chap. 6). Measurement capability needs to checked before embarking on production of measurement results.

Quality assurance of measurements analogous to assuring the quality of production with ISO 9000 (Chap. 1) is the focus of standard ISO 10012, alongside the ISO 17025 standard which deals with assuring the quality of testing and calibration laboratories.

Metrological confirmation is defined (ISO 10012 §3.5) as follows:

> 'Set of operations required to ensure that measuring equipment conforms to the requirements for its intended use.
>
> Note 1 Metrological confirmation generally includes calibration and verification, any necessary adjustment or repair, and subsequent recalibration, comparison with the metrological requirements for the intended use of the equipment as well as any required sealing and labelling.
>
> Note 2 Metrological confirmation is not achieved until and unless the fitness of the measuring equipment for the intended use has been demonstrated and documented.
>
> Note 3 The requirements for intended use include such considerations as range, resolution and maximum permissible errors.'

Additionally the following activities should be planned in general, although for once-off method use it is usually sufficient to perform metrological confirmation only (Table 4.1).

In the rest of this section about metrological confirmation, we focus on measurement uncertainty. The remaining processes of metrological conformity assessment will be assumed to follow by analogy the corresponding process of product conformity assessment, as treated in the remaining chapters of this book.

Table 4.1 Processes for quality assurance of measurement

Demonstration that a certain performance is fulfilled	Each performance in question should be in the first case one that the actual laboratory can potentially affect, while for other demands the results of validation and verification may suffice. Guidance may be sought for example in 'Proficiency testing by interlaboratory comparisons' (ISO/IEC 17043:2010)
Internal control	Continuous control of the total functioning of the measurement method with the help of control or check standards
External control	Participation in external control programs in order to investigate performance in relation to that of other laboratories, such as interlaboratory comparisons (Sect. 2.5.1)
Follow-up	Recurring analysis of control results in order to review measurement uncertainty assessments and to judge the need of and possibilities of improvement

4.2.1 Calibration and Metrological Confirmation. Uncertainty and Unknown Errors

According to the measurement management system standard ISO 10012 §7.2.2, measurement process design for metrological confirmation should include the identification and quantification of:

... Performance characteristics ... for intended use of measurement process...including:

- measurement uncertainty,
- maximum permissible error.

For our example of pre-packaged goods, Table 2.3 in Chap. 2 contains a number of requirements on measurement performance (in turn based on product specifications), such as the maximum permissible uncertainty as well as the maximum permissible error allowed for the instrument, irrespective of whether we are weighing a product or judging the perceived prestige of product. Measurement system specifications are summarised in Sect. 2.2.4.

In the present chapter concerning the actual performance of measurement, it is appropriate to give the following account of how to evaluate measurement uncertainties, in order to provide a quantitative basis for decision-making about metrological confirmation. These methods of uncertainty evaluation can also be used when summarising the quality of the actual measurements at hand, as will be exemplified here and in the case studies at the end of this chapter. Measurement uncertainties will need to be evaluated at each of the three steps and at every level shown in Fig. 4.1.

The task of evaluating measurement uncertainty is one of the most challenging since one is dealing with the unknown. One definition of measurement uncertainty is as an estimate of 'unknown measurement errors'. In most cases there are seldom time or resources to investigate all possible sources of measurement error. Where knowledge about measurement is limited—as it always is—measurement results have uncertainty.

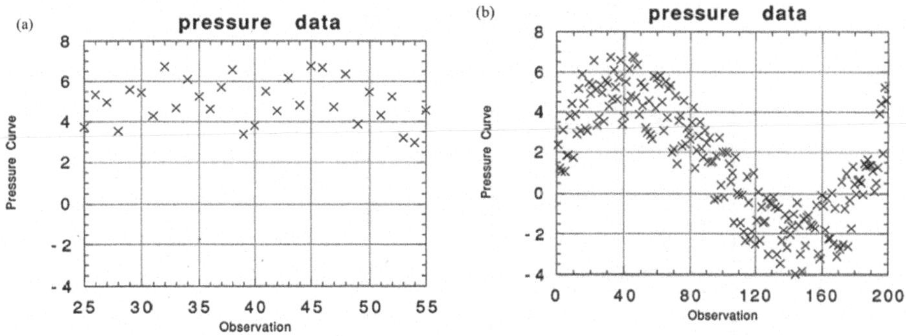

Fig. 4.2 Example of scatter in measurement data (**a**) short time interval; (**b**) whole day data

An illustrative example is shown in Fig. 4.2 which is real data from an automatic measurement system for atmospheric pressure: measurement scatter registered with a computer during a short time interval (about mid-morning, Fig. 4.2a) indicated a certain mean value and a scatter of about a couple of units of pressure (this was a high-performance barometer with a resolution of 0.5 Pa). A similar scatter was seen on inspection of the computer readout during a correspondingly short time interval during the afternoon, but unexpectedly the mean pressure appeared to have changed by several times the short-term scatter, despite little change in actual atmospheric conditions. Suspicions were confirmed when a plot (Fig. 4.2b) was made of the complete data collected by the computer that day. A final investigation revealed that the temperature regulation of the barometer sensor (bourdon tube) had malfunctioned, leading to a slow, sinusoidal swing in temperature—and consequently, false pressure reading—with a period of about 12 h. Other examples of the challenges of evaluating measurement uncertainty may be found in Sect. 2.5.2.

Measurement uncertainty evaluation requires a certain amount of statistics, but above all a considerable amount of knowledge and experience about measurement. For a more professional exposé of statistics than the brief survey made here, the reader is referred for example to Montgomery (1996).

Referring back to the picture of a measurement process shown in Fig. 4.1, the first steps (A) and (B), Calibration and testing of a measurement system, have been described in Sect. 3.1.2. It is typically at the third (C) implementation to measure stage that plots of the scatter of measurement results (Fig. 4.3) are made. The two dimensions of such plots are: on the x-axis is the quantity of interest while on the y-axis the height of the plot indicates the probability, p, of obtaining each quantity value. The resulting distribution of measurement values is termed a probability mass function (PMF) in the discrete, categorical case or a probability density function (PDF) for continuous variables. The height of the curve indicates initially the number (frequency) of observations at each measurement value. After observations cease, statistical thinking describes the corresponding probability of obtaining each value in the long run. Figure 4.3 is another way of presenting scatter in measurement data, compared with the 'bull's eye' approach shown in Fig. 2.2 or the time series approach of Fig. 4.2.

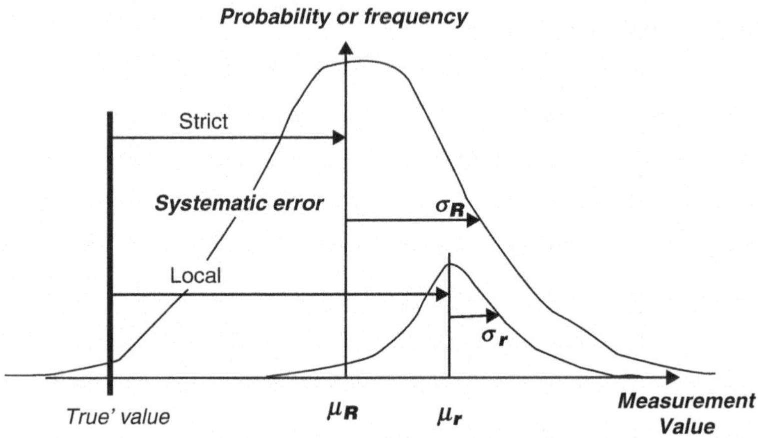

Fig. 4.3 Probability distributions (PDF) under conditions of repeatability (r) and reproducibility (R)

Repeatability and Reproducibility

It is important to state explicitly under which conditions the measurements have been made. Ideally an extensive series of observations should be made, not only under repeatability conditions, but also under reproducibility conditions (Sect. 2.2.2), in which each new measurement in principle is made with a different measurement system. In Fig. 4.3 are shown probability distributions typical for these two sets of conditions.

In most applications, one is interested not only in a locally valid metrological characterisation of the measurements, but also something more absolute, namely the reproducibility characteristic of more than one measurement set-up. Despite the wider spread of the measurement values under reproducibility conditions, as exemplified in Fig. 4.3, the mean μ_R will nevertheless generally be closer to the (unknown) true value than the mean μ_r under repeatability.

Convenience Sampling

A common question often raised at the start of a series of measurements is: how many measurements are needed to achieve the reliability required to test product adequately (Sect. 2.2.4)? If the measurements at hand are being performed for the first time one of course does not know what reliability will be achieved. In such cases it is common to perform 'convenience sampling', which consists of a limited number of measurements. Guidance for example in psychometric studies is to start with at least 20 instruments (that is, test persons) and/or 20 items (Kruyen et al. 2012). The reliability coefficient (R_z) for the item attributes, z, is calculated using Eq. (2.11) for the initial convenience sample. The Spearman–Brown prediction formula (Eq. (4.1), Spearman 1904, 1910), relates reliability coefficients, R_C and R_T, for the current (C) and target (T) tests to the corresponding test lengths, L_C and L_T, that is, the number of instruments (persons) or objects (items):

$$L_T = L_C \cdot \frac{R_T \cdot (1 - R_C)}{R_C \cdot (1 - R_T)} \tag{4.1}$$

As an example, consider an initial convenience sample of $L_C = 11$ test persons which turns out to have a reliability R_C of 0.57. Equation (4.1) readily yields the number of test persons, $L_T = 33$ persons needed to reach a reliability $R_T = 0.8$ recommended for high-stake testing where half of the observed scatter arises from measurement uncertainty (Linacre 1994).

Ultimately, fit-for-purpose measurement will mean designing experiments to balance the costs of measurement against the risks of uncertain measurement (Pendrill 2014, Sect. 2.1.1 and Chap. 6).

4.2.2 *Evaluating Measurement Uncertainty: Physical Measurements*

A series of n repeated measurements at a given nominal level of the measurand on a quantitative interval or ratio scale (Table 3.3) (e.g. mass 500 g or a task of a certain difficulty) will furnish a number of estimates, z_i, typically summarised statistically in terms of a mean value, $\bar{z}$, with an associated (standard) measurement uncertainty $u(\bar{z})$ equal to the standard deviation of the mean (JCGM 100:2008). Table 4.2 contains a summary of the formulae often used to calculate the measurement uncertainty of repeated measurements.

The formulae in Table 4.2 can be evaluated if one has resources to perform a number of repeated observations. With fewer resources, a number of more economical alternatives to a full, mathematical treatment are available, as follows. Standard measurement uncertainties of a few (up to 10) observations can be obtained as adequately trustworthy estimates by calculating the maximum variation width $=z_{max} - z_{min}$, and using:

Table 4.2 Example of Type-A evaluation of uncertainty of repeated measurements

Mean	$\bar{z} = \frac{1}{n} \cdot \sum_{i=1}^{n} z_i$
Standard deviation	$s = \sqrt{\frac{1}{n-1} \cdot \sum_{i=1}^{n} (\bar{z} - z_i)^2}$
Standard uncertainty = standard deviation of mean[a]	$u(\bar{z}) = s(\bar{z}) = \frac{s}{\sqrt{n}}$

[a]Regarding standard uncertainty instead as half a 67%-confidence interval leads some authors to expand this estimate by multiplying with the Student t-factor for the given number of degrees of freedom

1. the formula $s_{\text{Normal}} = \frac{Max}{d_n}$, where the factor d_n—given in Table 4.3—depends on the number, n, observations. This applies to cases where a Normal distribution can be associated with the measurement data (Reynolds 1987).
2. the formula $s_{\text{Rectangular}} = \frac{Max}{2 \cdot \sqrt{3}}$ applies to the rectangular (or uniform) distribution shown in Fig. 4.4. A typical example is a measurement situation where the resolution (Table 2.4) of a measurement instrument determines the scatter in the data. In that case, Max = resolution.
3. the formula $s_{\text{Triangular}} = \frac{Max}{2 \cdot \sqrt{6}}$ applies to the triangular distribution shown in Fig. 4.5. A typical example is a measurement situation where estimating ocularly the position of a pointer of a measurement instrument against a scale determines the scatter in the data.

Table 4.3 Values of factor d_n

n	2	3	4	5	6	7	8	9	10
d_n	1128	1693	205	2326	2534	2704	2847	297	3078

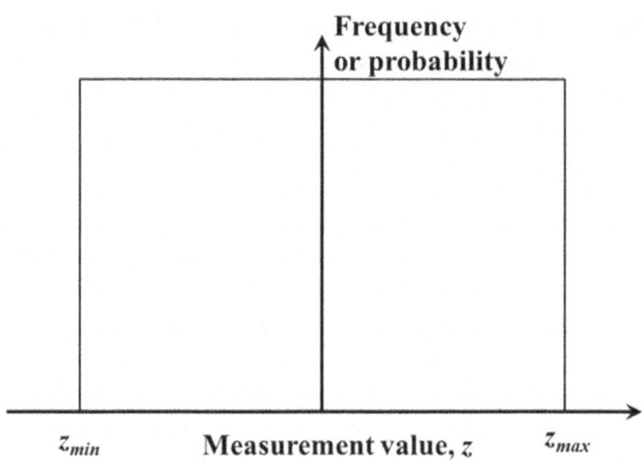

Fig. 4.4 Rectangular (or uniform) distribution

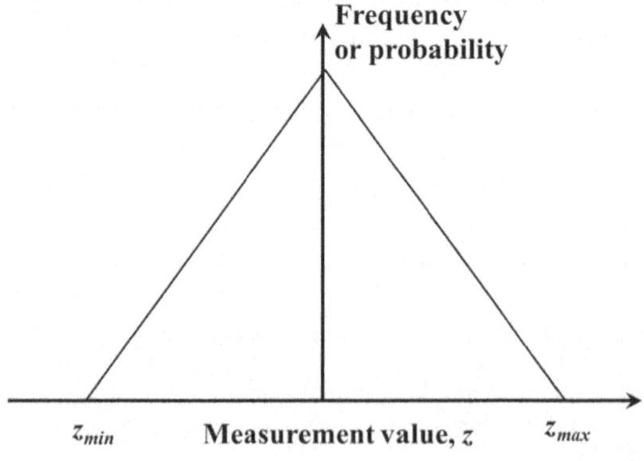

Fig. 4.5 Triangular distribution

4. the formula $s_{U\text{-shape}} = \frac{Max}{2 \cdot \sqrt{2}}$ applies to the U-shaped distribution shown in Fig. 4.6. A typical example is a measurement situation where an underlying sinusoidal noise signal, such as 50 Hz disturbances, determines the scatter in the data.
5. the formula $s_{\text{Trapezoid}} = \frac{Max}{2 \cdot \sqrt{2}}$ applies to the trapezoidal distribution shown in Fig. 4.7. A typical example is a measurement situation where one uses a measurement instrument which has been previously verified as performing within specifications, such as in legal metrology, where the display of the instrument, z_{IND} is found to satisfy the prescribed limits in terms of maximum permissible error, MPE, as described in Sect. 2.2.4. Because of measurement uncertainty, $U_{\max}$, when verifying (Sommer and Kochsiek 2002), the 'sharp' limits of the rectangular distribution (Fig. 4.3) become 'rounded' into the trapezoidal shape (Fig. 4.7).

Fig. 4.6 U-shaped distribution

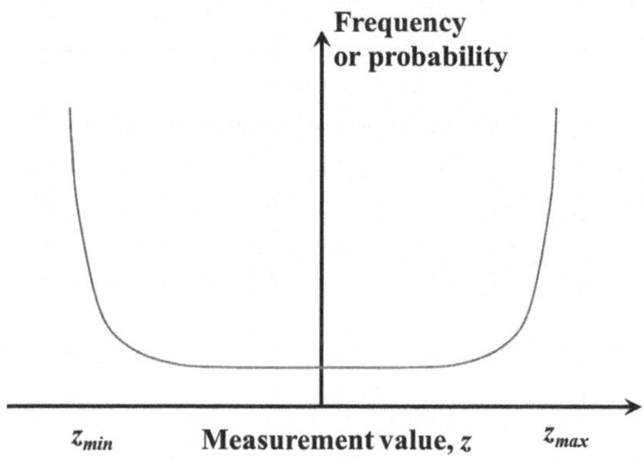

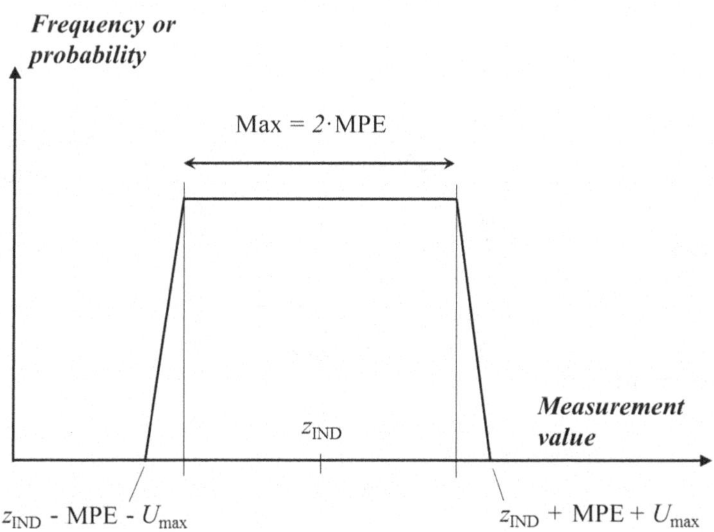

Fig. 4.7 Trapezoidal distribution (adapted from Sommer and Kochsiek (2002))

Apart from statistical estimates (type A) of uncertainty, based on repeated measurement, it is also recommended (JCGM 100:2008) to calculate uncertainties using prior knowledge (type B), such as (a) deriving a calibration uncertainty by dividing the expanded uncertainty, U_{cal}, by its coverage factor, k_{cal}, when an instrument or standard (etalon) was calibrated, or (b) using instrument specifications (such as in Fig. 4.7). Depending on the required accuracy, specifications by the instrument manufacturer based on testing of many instruments may in some cases be too 'wide', and it will normally be worthwhile to have the actual instrument at hand calibrated individually.

The mean value, $\bar{z}$, with an associated (standard) measurement uncertainty $u(\bar{z})$ of any series of measurements will in general be the result of how measurement information is propagated through the measurement system including the restitution from the response, y, as described in Sect. 2.4; the case studies in Sect. 2.7; and again in Chap. 5. A visualisation of this transmission of measurement information can be made by drawing an Ishikawa ('fishbone' or cause-and-effect) diagram (Fig. 1.5 in Chap. 1), in the specific case of a measurement system, as exemplified in Fig. 4.8, which can be useful when setting up a budget over all conceivable sources of measurement uncertainty.

Each element of the measurement system (object, instrument, method, operator, etc. (Fig. 2.4)) is depicted with one major 'bone' of the Ishikawa diagram. In some cases, the uncertainties propagating through an 'instrument' can be further modelled in terms of a chain of elements representing sensor; signal conversion; signal conditioning; data conversion (Eq. (2.5), sect. 2.4.3, Bentley 2004; Sommer and Siebert 2006).

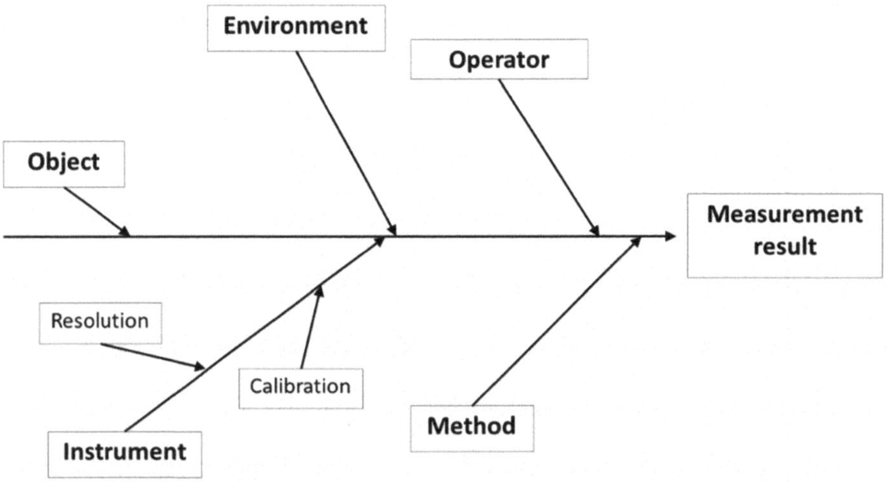

Fig. 4.8 Ishikawa diagram based on MSA

Mathematically, measurement uncertainty propagates as standard deviations, σ, or more readily as the addition of variances,[1] thus following for example (see Sect. 2.4 for an explanation of the notation):

$$\sigma_{O_j}^2 = \sigma_{I_{j+1}}^2 = \left(\frac{\partial O_j}{\partial I_j}\right)^2 \cdot \sigma_{I_j}^2 + \left(\frac{\partial O_j}{\partial I_{M_j}}\right)^2 \cdot \sigma_{I_{M_j}}^2 + \ldots \tag{2.7}$$

where the partial derivatives are called sensitivity[2] coefficients (JCGM 100:2008).

Similar expressions can be formulated for any measurement model, for instance the response output, y, of an interlaboratory experiment Eq. (2.12).

The final measurement result is of course not the output response, $y = R$, of the measurement system but the estimate of the measurand (stimulus attribute $S = z$), in which case the uncertainty in the item attribute, z, can be derived from the restitution formula $z_j = S_j = \left(\frac{R-b}{K}\right)_j = \frac{y_j - b_j}{K_j}$ Eq. (2.9) as:

$$u(z = S) = \mathrm{u}\left(\frac{R-b}{K_{\mathrm{cal}}}\right)$$
$$= \frac{R-b}{K_{\mathrm{cal}}} \cdot \sqrt{\frac{u(R-b)^2}{(R-b)^2} + \frac{u(K_{\mathrm{cal}})^2}{K_{\mathrm{cal}}{}^2} - 2 \cdot \rho \cdot \frac{u(R-b) \cdot u(K_{\mathrm{cal}})}{(R-b) \cdot \mathrm{K}_{\mathrm{cal}}}} \tag{4.2}$$

where ρ is the degree of correlation between $(R - b)$ and K_{cal} (Hannig et al. 2003).

Care has to be taken when evaluating Eqs. (2.7) and (4.2) that uncertainty components—such as when calibrating instrument sensitivity—are not counted twice. An example of the evaluation of Eq. (4.2) is given in the case study of pre-packages at the end of this chapter (Sect. 4.5.1).

Treatment of uncertainties in the restitution process in the particular case where, as in the social sciences, Man acts a measurement instrument will be examined later (Sect. 5.4.1).

Summarising the steps to be taken overall when expressing measurement uncertainty as described in this section (JCGM 100:2008):

1. Analyse the measurement system. Set up an error budget.
2. Correct for known measurement errors, such as subtracting for known bias, b, and sensitivity K differing from 1 (unity).
3. Evaluate (standard) measurement uncertainties with methods of type A alternatively type B.
4. Combine standard measurement uncertainties by quadratic addition (e.g. Eq. (2.7)) $\geq u_c$.
5. Expand measurement uncertainty $\geq U = k \cdot u_c$.

[1]Plus eventual correlation terms among the different sources of uncertainty.

[2]Not to be confused with instrument sensitivity K.

4.3 Accurate Measurement across the Disciplines

The extension of metrology towards the social sciences is the latest in a long series of widening the scope of quality-assured measurement to encompass engineering, physics, chemistry and other fields. The next sections give a brief review of this process.

4.3.1 Accurate Measurement: Is It the Domain of the Engineer or the Physicist?

> The number of electrical measuring instruments recently devised is very great. The practical man is not satisfied with the delicate instruments of the physicist, whilst the latter, of course, cannot be satisfied with the results of the measuring instruments arranged by engineers and technical electricians, however satisfactory for industrial purposes (The Telegraphic Journal 1884, quoted in (Gooday 1995))

Measurement accuracy is an elusive concept, often with different meanings for different people as illustrated by the above quote from over 130 years ago. Using the terminology of international standardisation (ISO 5725) where accuracy is defined in terms of both precision (amount of scatter in repeated measurement data) and trueness (size of systematic error), one may observe the following:

A broad generalisation would be to assign the task of achieving good precision to the measurement engineer, whereas it is the task of the physicist to provide best estimates of the true value of a physical quantity. Here 'truth' refers not simply to a freedom from error, but to something rather more absolute (see further in Chap. 3).

While physicists have clearly had a major influence on the formulation of metrology in the past 100 years or so, a more engineering—'black box' (such as a classic VAOmeter)—approach, while robust, admittedly has led to significant omissions, such as not emphasising enough that measurement is a 'concatenation of observation and restitution' (Fig. 3.3). Even analytical chemists risk making a similar mistake since modern instrumentation, such as for digital polymerase chain reaction (PCR) quantification of DNA, are very much of black box, robust character rather than visualised as instruments built bottom-up, from first principles, where delicacy is traded against insight.

4.3.2 Metrology in Physics and Chemistry

In meeting demands for quality assurance, what is common ground and what is different between the views of metrological traceability of the chemist and the physicist—to take just two disciplines? The present attempts at spanning the physical and social sciences can thus be seen as a continuation of the 'bridging of the cultural gap' between metrologists and analytical chemists mentioned a number of years ago by King (1997).

What basic differences are there between the provision of a metrological infrastructure for physical measurements as opposed to (bio)chemical measurements (Pendrill 2005)? Certainly, the amount of substance is a typical measurement quantity in chemistry and confirming the analyte's identity is a distinguishing feature of measurements in chemistry.

'Not Always Fully Traceable to the SI': Only for the Chemist?

Reference materials (some of which are certified, CRM) are used perhaps more extensively in measurements in chemistry than in physics. In some cases these offer practical solutions in providing reference values where full traceability to the SI is not easily realised. Use of CRMs is not, however, exclusive to the chemist, and physical CRMs for quantities such as material hardness or the optical brightness of paper are used widely. Pradel et al. (2003) indicate that CRMs are used in 14% of applications of a physical nature. The European Commission's Bureau of Certified Reference Materials indeed produced its first CRMs in the physical area (BCR 1985).

In cases where traceability is provided in an unbroken chain of comparisons to the SI, then not only can the results of measurements of a particular quantity be compared, but also measurements of different quantities thanks to the coherence of physical laws (Sect. 3.4.1). Such coherence cannot however generally be assumed when obtaining traceability through the use of CRMs.

Little Interest in Traceability?

> ... end-user chemists ... no philosophical interest in knowing 'if a bias must always be corrected'... (Charlet and Marschal 2003)

There are measurement situations admittedly where metrological traceability is less important, perhaps more commonly for the chemist and in testing than in physics (Mittmann et al. 2003). This can be where local conditions of measurement, such as environmental factors, the choice of measurement method and the skill of individual operators, can affect the measurement result much more than any, perhaps small, uncertainty associated with an uncorrected bias with respect to reference values. In such cases, the lack of full traceability to the SI typical when using CRMs described above will maybe not be a serious problem.

King (2003) has stated this succinctly: 'The essential requirement is that ... traceability is ... established at a level of uncertainty appropriate to the final test result.'

A pragmatic recommendation is that calibration uncertainties should not exceed half of the total measurement uncertainty, as an appropriate quantitative limit for when metrological traceability is significant in testing and measurement (Pendrill 2005, Sect. 3.4.3).

Quantitative rules specifying uncertainty limits of this kind are now becoming established, for instance, in legal metrology and in geometrical product specifications, which set a limiting value on the total (usually expanded) measurement uncertainty in relation to the tolerance in the context of testing for conformity assessment (Källgren et al. 2003)—for instance, several of the measuring instruments covered by the EU Measurement Instrument Directive (2014) are to be tested according to OIML recommendations where the expanded uncertainty (with coverage factor $k = 2$) on testing should not be larger than one-third of the tolerance.

Can the Physicist Learn Anything About Metrological Traceability from the Chemist?

The correct description of the propagation of measurement errors in any actual measurement situation, be it in a well-controlled laboratory environment or in the field, is as necessary in dealing appropriately with metrological traceability as in estimating measurement uncertainty.

Often the same measurement model (Fig. 2.4, Sect. 2.4.1), expressed as a functional relation connecting input and influence quantities to the measurand, can be invoked as a prelude to treating both traceability and uncertainties (EURACHEM/CITAC 2003). As emphasised in the EURACHEM guide, for example, an important requirement is validation (Sect. 2.5)—that is, that the measurement method and other components included in the measurement model are working properly—which needs to be demonstrated in addition to considering traceability. Formulation of the measurement model is arguably the most difficult step in any traceability or uncertainty investigation, since it requires a considerable insight into the measurement situation. No amount of statistics will fully compensate for the effects of a missed component in the measurement model (Sect. 2.2.3). The propagation of metrological traceability as well as the measurement uncertainty arising from all significant stages in the measurement procedure need, in general, to be considered. As stated by Golze (2003), '. . . for traceability . . . (in) chemical analysis and testing, the underlying concepts have to be generalised'.

In the physical calibration laboratory, with its well-controlled environment, standard measurement methods, trained operators and (very) homogeneous samples substantially free of matrix effects, measurement uncertainty evaluation often focuses on only certain types of sources of uncertainty, such as those arising from the measuring instrument (see further discussion, for example, in EA-4/16 (2003)). Bearing in mind the experiences of metrologists in chemistry and testing laboratories recounted above, perhaps a more appropriate approach to treatment of metrological traceability, as well as measurement uncertainty, would be to employ the concept of measurement system analysis (MSA) as described in Chap. 2 of this book.

The process of clarifying and implementing basic concepts of metrological traceability in chemistry might lead to increased insight even for the physicist. Paradoxically, while metrological traceability is well known to the laboratory physicist, ensuring the quality and comparability of measurements in complex industrial and field situations far from the well-controlled laboratory environment is increasingly demanding renewed insight into the traceability concept even for the experienced applied physicist and engineer. One conclusion is that there is much common ground and opportunity for mutual exchange of views and insights between the chemist and the physicist in questions of metrological traceability (Pendrill 2005).

4.3.3 Metrology in the Social Sciences

In comparing measurement in the physical and social sciences, Hand (2016) claims that the former often involve representational measurement—i.e. that the measurement object is represented by a number—while in the social sciences, pragmatic measurement is more often used, that is a measure is constructed to have the 'right sort of properties for our intended use'. Our discussion in Chaps. 2 and 3 about design of experiment illustrate the importance in all kinds of measurement—even in the physical sciences—of setting up your measurements in a fit-for-purpose and pragmatic way, by vectoring in from the start the purpose of measuring a particular object. So a simple division of concepts in terms of representational contra pragmatic measurement does not seem viable. Rossi (2014) has made a similar point, mentioning the valuable contribution of Finkelstein to uniting metrologists in the physical and social sciences in describing measurement as 'a process of associating numbers, in an empirical and objective way, to the characteristics of objects and events of the real world in a way so as to relate them'. Table 3.1 indicates the breadth of our approach to measurement throughout the physical and social sciences.

4.4 Metrological Concepts in Social Science Measurements

A key concept according to our approach to measurement in the social sciences is Man as a Measurement Instrument, as described in Sects. 1.2.2 and 2.4.5 and as follows.

As we have seen above, much of the established metrological terminology and concepts (Sect. 4.2.2) carry well over into measurement in the human sciences. One major caveat however is, as mentioned, that data obtained from the response of a measurement system where Man is the instrument are often not amenable to the usual statistical tools (e.g. calculating a mean and standard deviation) (3.5.1). The recommendation is that one instead transforms the measurement system response $P_{success}$ (e.g. the probability of making a correct decision or performing a task of a certain difficulty) lying on a less quantitative ordinal or nominal scale (Table 3.3), onto a more quantitative interval or ratio scale by applying the psychometric Rasch (1961) formulation Eq. (1.1). This transformation is the core action of the restitution of the measurand, as described in Sect. 2.4.5 and in Fig. 3.3. After restitution, the transformed data—with estimates of the item δ and person (or probe) θ attributes—can be expected to lie on a more quantitative interval or ratio scale where most statistical tools and objective decision-making can be performed more reliably than on the 'raw' measurement system response P_{success} data. (Sect. 5.7 presents appropriate tests of the validity of the transformation.)

In general, on examination it will be found that the measured person attribute θ (e.g. a level of ability of a particular person or instrument) differs, because of limited measurement reliability, from the 'true' θ', with an error ε_θ:

$$\theta = \theta' + \varepsilon_\theta$$

and the limited reliability expressed in Eq. (3.1) evident in estimates of task difficulty: $\delta = \delta' + \varepsilon_\delta$. Such deviations arise at least in part because the measurement system (human being) used to 'probe' the object or item is not perfect (Sect. 2.1.2 and Fig. 2.3).

Wright (1994) described how the Rasch approach makes separate estimates of attributes of each test person (TP) i with attribute (e.g. ability) θ_i and of each item j with attribute (e.g. difficulty) δ_j. These two parameters are adjusted in a logistic regression of Eq. (1.1): $\theta - \delta = \log\left(\frac{P_{\text{success}}}{1-P_{\text{success}}}\right)$, to the score response data $y_{i,j}$ on an ordered category scale by minimising the sum of the squared differences:

$$\sum_{i=1}^{N_{\text{TP}}} \sum_{j=1}^{L} \left(y_{i,j} - P_{\text{success},i,j}\right)^2$$

The goodness of fit can be judged by examining how closely the overall fitted ogive item response curve of Eq. (1.1) matches individual average scores at different locations across the scale.[3]

Estimates of measurement uncertainty, u, for person (i, N_{TP}) and item (j, L) attributes and categories k, k', derived from the Rasch expression Eq. (1.1) are made, respectively with the following expressions (Wright 1994):

[3]In principle since the raw response data do not in general lie on a quantitative interval scale, these differences have no meaning. But the assumption is that only small differences between observed and fitted values are examined, in which case uncertainties in scaling are in most cases negligibly small. See chap. 5 for more discussion.

$$\begin{cases} u(\theta) = \sum_{j=1}^{L} \left(P_{\text{success},\, i,j,k} \cdot P_{\text{success},\, i,j,k\prime}\right)^{-\frac{1}{2}} \\ u(\delta) = \sum_{i=1}^{N_{\text{TP}}} \left(P_{\text{success},\, i,j,k} \cdot P_{\text{success},\, i,j,k\prime}\right)^{-\frac{1}{2}} \end{cases} \tag{4.3}$$

The dichotomous relations of the basic Rasch model can be extended to the multinomial, polytomous case, according to the expression:

$$q_{i,j,c} = P\left(y_{i,j} = c\right) = \frac{e^{\left[c \cdot (\theta_i - \delta_j) - \sum_{k=1}^{c} \tau_{k,j}\right]}}{\sum_{c=0}^{K_j} e^{\left[c \cdot (\theta_i - \delta_j) - \sum_{k=1}^{c} \tau_{k,j}\right]}} \tag{4.4}$$

The polytomous Rasch variant of GLM then models the response Y at any one point on the scale as a sum of dichotomous functions expressed as the log-odds ratio $z = \theta - \delta + \tau$ for each threshold τ, where the latter is defined as the point on the entity value scale, z, where the probabilities of scoring are equal at 50%. This polytomous Rasch variant is referred in the literature as the Andrich (1978) 'rating scale', Masters (1982) 'partial credit' approaches, among others. In *R* this is referred to as *Extended Rasch Modeling: The R Package eRm*. Programs such as WINSTEPS make a logistic regression of the polytomous Rasch formula to the response data $Y = P_{\text{success}}$, using the 'Joint Maximum Likelihood Estimation' method to estimate values of the 'latent' (explanatory or covariate) variables Z: θ, δ and the thresholds τ.

Naturally, any assumptions—such as about invariance and dependency—need to be tested quantitatively in any specific formulation of the Rasch model to a set of data. Linacre and Fisher (2012) write: 'An advantage of Rasch methodology is that detailed analysis of Rasch residuals provides a means whereby subtle inter-item dependencies can be investigated. If inter-item dependencies are so strong that they are noticeably biasing the measures, then Rasch methodology supports various remedies.' Further discussion of the validity of the Rasch model can be found in Chap. 5.

The next section illustrates the treatment of uncertainty by examining an elementary counting task as a prototype for measurements in the social sciences and more widely.

4.4.1 Elementary Counting

The ability of a person to perform an elementary task, such as counting the number of dots, can be readily judged by displaying sets of dots—such as shown in Fig. 4.9—with successively increasing numbers of dots. Such a task, although trivial

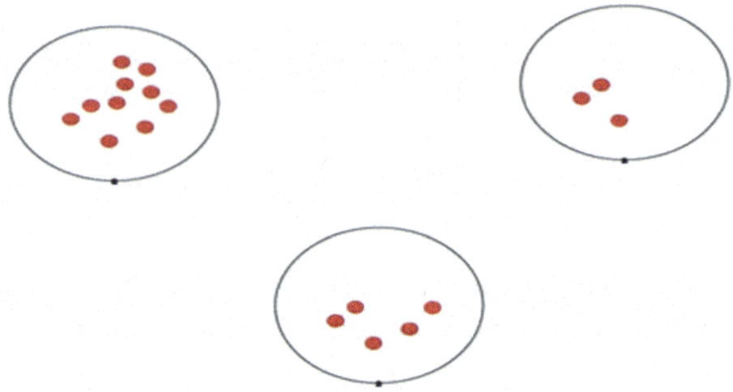

Fig. 4.9 How many dots?

for an educated adult, can be challenging for the higher numbers of dots if each set is displayed only a couple of seconds: look quickly at the leftmost figure—are there 8, 9, 10 or perhaps 11 dots? For kindergarten children and for certain tribes, such as the Munducuru Indians in the Amazon jungle, such tasks are equally challenging even if infinite time is given to inspect each set of dots (Pendrill and Fisher 2015).

A measured count, Y_k, (response of the measurement system) is given by:

$$Y_k \cdot [I] = \{M + k \cdot \varepsilon\} \cdot [I]$$

where ε is the resolution step separating discrete category levels on the horizontal axis and $[I]$ denotes the 'unit' of counting, i.e. one 'dot' in the present case. (Sect. 3.6.1 has further counting examples.)

Underestimation can lead to a spread in the estimated classifications over a number of lower discrete levels (Fig. 4.10), where each classification error is a certain integer multiple, k, of ε, and overestimation to a corresponding spread over a number of higher levels, in both cases compared with the 'true' level, M. For simple integer counting, naturally ε equals unity; for other tasks, the discrete resolution of the human 'instrument' will set the least discernible difference, ε, between two adjacent category levels.

The example of the Munducuru Indians can serve as a template for a wide range of qualitative measurement in the social sciences and elsewhere using in particular the Man as a measurement instrument model. In studies of elementary tasks—such as counting (Pendrill and Fisher 2015) or ellipticity perception (Pendrill 2013), an initial measurand could be, respectively, the number or degree of correlation in a cloud of dots. Man, in this case acting as a measurement instrument, yields estimates of the value of each measurand. But what is interesting is not the number or ellipticity of the clouds of dots since we know these already, but rather how well these measurements are performed. The measurand of interest is thus the ability to perform such measurements, described in terms of a decision-making process, e.g. how well can the human instrument resolve the difference between adjacent stimuli, such as: 'Are there nine or ten dots in the cloud?' This decision-making ability can be expressed as the probability of success, $P_{success}$, vis-à-vis the risks of making incorrect decisions.

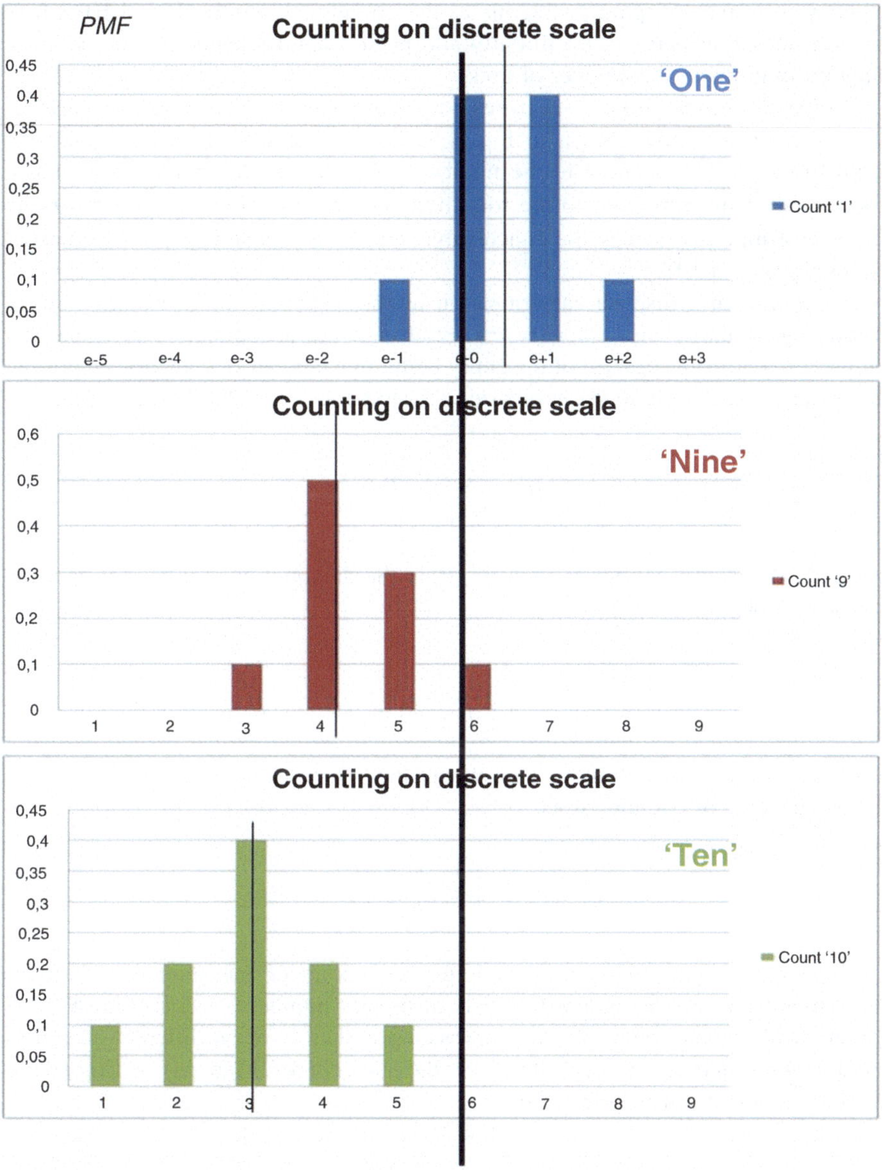

Fig. 4.10 Probability mass functions [PMF] for three cases of counting of increasing difficulty

A common questionnaire might have five categories of response—from completely disagree to completely agree. In such a case, the quantity of interest (the 'construct') underlies implicitly the response scale plotted on the 'horizontal axis' of the questionnaire. Questions about satisfaction for instance (or 'prestige', see case study 4.5.3) are rated with scores which with a proper Rasch decomposition (Sect. 4.4) yield separate estimates of product quality and person (instrument) leniency in a manner quite analogous to the counting example (Fig. 4.10) where the corresponding pair—task difficulty and person ability—underlies the label or category in terms of the number of dots.

N repeated observations ('measures' or 'classifications') are made of each set of dots in the example shown in Fig. 4.9. The probability, q_k, of assigning the count to category k is equal to the number, N_k, of times the count k is obtained ('occupancy') in proportion to the total number, N, of counts: $q_{k=}\frac{N_j}{N}$. Probability 'mass' distributions display the conceivable spread of probability q_k for perceived values, Y_k (Fig. 4.10).

In the case of a discrete categorisation of the response, the probability, q_c, of classifying an entity in a category c is related to the accumulation of (generally unobserved) probabilities, p_k, of the entity being in a number of categories k prior to classification according to the expression (Bashkansky et al. 2007, Sect. 2.5.3):

$$q_c = \sum_{k=1}^{K} p_k \cdot P_{c,k}$$

where the decision ('confusion') matrix, **P**, in the dichotomous (go/no-go) decision case would be:

$$P = \begin{pmatrix} 1-\alpha & \alpha \\ \beta & 1-\beta \end{pmatrix}$$

In the simplest, dichotomous case where the prior is known to be M, as in the elementary case of counting dots,

$$p_k = \begin{cases} 1 & k = M \\ 0 & k \neq M \end{cases}$$

If we regard such elementary counting as a kind of measurement, then the increasing errors in counting with levels of greater numbers of dots (Fig. 4.10)—perhaps with a distribution of errors across a population, or on different occasions for a certain individual—will illustrate the concept of Man as an (imperfect) measurement instrument (Fig. 1.1b).

4.4.2 Entropy, Perception and Decision-Making

Considering the decision-making process as part of the transmission of information in a measurement system as a perception of pairwise discrimination of adjacent stimuli (Fig. 6.3), such as in choice in cognitive psychology (Iverson and Luce 1998, Eq. (2.6)), the subjective (Kullback and Leibler (1951)) distance $D_{\mathrm{KL}}(a, b)$ between two stimuli, a and $b > a$, is expressed (Dzhafarov 2011) as the integral over the level, s, of stimulus of a measure of the ability to perceive a dissimilarity

$$P(s, s + ds) = Pr\ [b \text{ is judged to be greater than } a] :$$

$$D_{\mathrm{KL}}(a, b) = \int_a^b \frac{P(s, s + ds)}{ds} \cdot ds$$

The subjective distance, $D(a,b)$, reduces to Fechner's law used widely in psychophysics:

$$D(a, b) = k \cdot \log\left(\frac{b}{a}\right)$$

when the gradient of the dissimilarity $\frac{D(s, s+ds)}{ds} = \frac{k}{s}$ and a is set to the 'absolute threshold'.

This approach of relating subjective distance to accumulated dissimilarity in terms of discrimination probabilities can be extended to include not only continua of the senses (of colours, sounds, etc.) but also to Fechnerian scaling of the perception of discrete object sets where stimulus sets are 'isolated entities', such as schematic faces; letters of the alphabet; dialects and the like (Nerbonne et al. 1999).

In the simplest, dichotomous case where the prior is known to be M dots, as in the elementary case of counting (Fig. 4.9), the subjective distance $D_{\mathrm{KL}}(a, b)$ between the distributions (P) and (Q) for the two stimuli a and b to the measurement-based decision is obtained by substituting $\frac{P(s, s+ds)}{ds} = dP_{\mathrm{success}}$ and $ds = -z$ to yield:

$$\begin{aligned} D_{\mathrm{KL}}(P, Q) &= \int -z \cdot dP_{\mathrm{success}} \\ &= -[P_{\mathrm{success}} \cdot \log(P_{\mathrm{success}}) + (1 - P_{\mathrm{success}}) \cdot \log(1 - P_{\mathrm{success}})] \\ &= H(P, Q) - H(P) = H(Q|P) \end{aligned} \tag{4.5}$$

where we set an equivalence between the subjective distance $D_{\mathrm{KL}}(a, b)$ and the conditional entropy $H(Q|P)$. How measurement information is acquired, transmitted, lost and distorted on transmission through the measurement system and in communication more generally can be described in terms of entropy (both for measurement units and uncertainty) as discussed in 3.2.3.

A straightforward derivation of decision probabilities can be made with the method of Lagrange multipliers subject to the constraint of maximising entropy (Sect. 5.5.3), leading readily (Linacre 2006) to the logistic regression link function:

$$z = \log\left[\frac{P_{\mathrm{success}}}{1 - P_{\mathrm{success}}}\right]$$

Two examples demonstrate the breadth of application of these concepts of information entropy, uncertainty and decision-making—(1) counting dots and (2) comprehending language. As discussed more in Chap. 5, the information theoretical entropy, which is a measure of the amount of information, *I*, in messages, can be expressed in the simplest case of a number, *G*, of categories (Brillouin 1962) as:

$$I \sim \ln P \sim \ln(G!)$$

The larger the categorical multiplicity (*G*), the less the amount of order and the greater the task difficulty, δ.

1. For the Mundurucu Indians, the difficulty, δ, of counting increasing numbers (*G*) of dots (Fig. 4.9) in fact is found to progress, as expected, as the logarithm of that number, in line with the entropy (as a measure of order). The same dependence of task difficulty on category multiplicity can be found in many elementary tasks, such the number of blocks, digits, words and the like to be remembered in sequential tests in cognitive assessment. The degree of order and consequently the level of difficulty experienced when attempting to make organisations (hospitals, factories, etc.) more efficient can also be modelled in terms of informational entropy based on category multiplicity (simply the number of options available).
2. The well-known asymmetries in the mutual comprehension of the three Scandinavian languages can be modelled in terms of differences in the conditional entropy $H(Q|P)$ reflecting different categorical multiplicities of the various languages. Quoting Moberg et al. (2007): 'As a simplest illustration of how conditional entropy can be used for linguistic units, consider the following. Written Danish words have only one vowel in their grammatical endings, the letter *e*, while Swedish uses *e*, *a* and *o*. This means that a Swedish speaker that encounters the Danish letter *e* has three options when trying to find the equivalent Swedish phoneme. Idealizing now to the situation where this were the only use of the sounds in question, we can see that a Danish speaker, upon encountering Swedish *e*, *a* or *o*, can know that the proper correspondence is *e*. The entropy is therefore higher for Swedish given Danish in this example, and the relationship is asymmetric'.

In our research on cognitive tests for people with neurodegenerative diseases, both the difficulty of each cognitive task (such as remembering sequences—tapped blocks, numerals, words, etc.) and the cognitive ability of each person are described in terms of entropy—the easiest tasks or most able persons have the least entropy (Pendrill 2018; Cano et al. 2019).

In the field of perception, two principles—of likelihood and simplicity—have been distinguished as follows (van der Helm 2000; Pizlo 2016):

- The likelihood principle infers the probability of an interpretation of a perception from an analysis of the world (i.e. the entity).

- The simplicity principle infers this probability from an analysis of the interpretation itself (as perceived by Man as a measurement instrument in our words).

Although rarely done in perception studies, psychometric (Rasch) analysis is in our opinion a method of choice when making separate estimates of for instance (1) the ability of each human instrument (or other probe) and (2) the inherent level of challenge posed by a particular object or task. Historically attention has indeed shifted (for instance in psychology) back and forth between a focus on the ability of a person or on a stimulus from an object (such as a task). In considering the symmetry in the person and item perspectives on construct validation in the Rasch approach, Stenner and Smith (1982) quote the following text by Thurstone (1923), p. 364:

> I suggest that we dethrone the stimulus. He is only nominally the ruler of psychology. The real ruler of the domain which psychology studies is the individual and his motives, desires, wants, ambitions, cravings, and aspirations. The stimulus is merely the more or less accidental fact...

Stenner and Smith (1982) found it 'baffling that early correlationalists were so successful in focusing the attention of psychometricians on person variation'.

The present approach in a sense shifts focus once again to the individual—who is identified as the measurement instrument—rather than the questionnaire item commonly referred to as an 'instrument' (Sect. 1.2.2).

These ideas are considered more broadly in Chap. 5, for instance considering the concept of 'prägnanz' adopted by so-called Gestalt psychologists in describing perceptual simplicity in general terms (Koffka 1935). There seem to be extensive applications of entropy, perception and decision-making in a wide variety of contexts in explaining item and person attributes when assuring the quality of categorical data on ordinal and nominal scales.

4.4.3 Construct Specification Equations

The above discussion, particularly how the concept of entropy can explain the item δ and person (or probe) θ attribute values obtained, opens up possibilities of formulating a construct specification equation (mentioned in the 'product' description Sect. 1.4 and in Sect. 3.3.2) for these attributes. This will be an expression relating the Rasch construct $\boldsymbol{Y}$ (e.g. task difficulty, δ, or person ability, θ) linearly to a set of explanatory variables X:

$$\hat{Y} = \sum_k \beta_k \cdot X_k \tag{4.6}$$

Apart from the mean, $\hat{\theta}$, and standard uncertainty, $u(\theta)$, of the attribute value for each person (or δ for an item attribute), it is of interest to express corresponding means and standard uncertainties in the regression coefficients, β, of the construct specification equation relating each attribute value to a set of explanatory variables ('manifest predictors', $\boldsymbol{X}$) (mentioned in the 'product' description Sect. 1.4), as well as statistics for significance testing of various differences among attribute values. A construct specification Eq. (4.6) expresses a theory regarding observed regularity in a set of observations generated by a measurement procedure for a set of objects (e.g. tasks).

The specification equation approach as a means of formulating 'recipes for certified reference materials' in the social sciences can be applied to cognitive task difficulty with multimodal statistical approaches used to correlate cognition and function outcome scores with test sequence entropy, length, etc., thus providing a predictive tool for the design of new cognitive tasks complementing existing scales and demonstrating item equivalence. The case study given at the end of this chapter—Sect. 4.5.2—includes an account of formulating a construct specification for product 'prestige'. Further details are given in Chap. 5.

4.4.4 Uncertainty and Ability

A couple of instances of uncertainty from different elements of the measurement system can be cited:

- Uncertainty in the stimulus $u(z)$ in the link function $z = \theta - \delta$, of whatever source, will propagate through the measurement system, giving an uncertainty in the response equal to $K \cdot u(S) = \frac{\partial P_{success}}{\partial z} \cdot u(z)$. This expression is investigated in depth in Chap. 5 where it is shown to lead to characteristic dispersion symptomatic of person-dependent scaling in plots such as construct alleys of fit residuals in the response.
- A second component of uncertainty is associated with instrument sensitivity $u(K_{cal})$. Since this uncertainty appears in the response, it will also propagate through to the above uncertainties, $u(z)$, in the estimate of the measurand (i.e. the stimulus), via the restitution process. Uncertainties in the instrument sensitivity $u(K)$ in the physical sciences will contribute:

$$u(z = S) = u\left(\frac{R - b}{K}\right) = \frac{R - b}{K^2} \cdot u(K) \tag{4.7}$$

for the relative standard uncertainty in the item attribute value according to Eq. (4.3).

The same arguments used by Humphry (2011, Chap. 3 Eq. (3.7)) to introduce measurement units into the Rasch model without losing the all-important possibility of estimating person and item attribute values separately (which is not in general possible with the 3PL model) can also be used to introduce the instrument sensitivity, K, into the 'logistic measurement function', as described in Sect. 2.4.5. In most applications of the Rasch model it is assumed (and subsequently tested with fit statistics) that the sensitivity (aka discrimination) is the same for all test persons. (An example of where this is not the case will be presented in Chap. 5).

For the present purpose, we therefore focus on the effects of uncertainties *u(K)* in the instrument sensitivity. By differentiating Eq. (2.11) one finds: $u(z = S) = -\frac{u(K)}{\rho}$. By comparison with Eq. (4.7), it can be seen that the discrimination $\rho = K^2$, i.e. the squared instrument sensitivity.

Conceptually when treating the ability of the 'instrument'—i.e. the person—at the heart of the measurement system as a significant factor in limiting measurement reliability, it is important to distinguish between measures of how well a human being (or other probe) in a measurement:

1. performs as a measurement instrument, in terms of uncertainties in the instrument sensitivity, $u(K)$,
2. performs tasks of a given level of difficulty, as measured in terms of Rasch attribute θ as a measure of ability.

There may well be a connexion: a more able person will probably function more reliably as a measurement instrument, but the two concepts are distinct. We defer further discussion to Chap. 5.

4.4.5 Separating Object and Instrument Attribute Estimation in Qualitative Measurement

Many qualitative measurements are still today made without the essential separation of instrument (probe) and measurement object variations (Sect. 2.2.3), thereby considerably compromising accuracy particularly since such measurements often involve rather 'noisy' measurement instruments, such as human beings.

In the words of Guilford (1936):

> It must be granted that, to measure such psychological attributes as appreciation of beauty…, we must depend upon secondary signs of these attributes. The secondary signs bear some functional relationship to the thing we wish to measure, just as the movement of a pointer on a scale is assumed to bear a functional relationship to the physical phenomenon under consideration. The functional relationship may be simpler and more dependable in the latter case than in the former and the type of relationship may be more obvious. That is the only logical difference. It is admittedly a difference of some practical consequence. But it is not a difference which leads to the conclusion that measurement is possible in the one case and impossible in the other.

$$\theta - \delta = \log\left(\frac{P_{\text{success}}}{1 - P_{\text{success}}}\right) \tag{1.1}$$

Note that, although Rasch's original formulation referred to human measurements, there is nothing in the mathematics of Eq. (1.1) which explicitly restricts the expression to cases where Man acts as a measurement instrument. As indicated

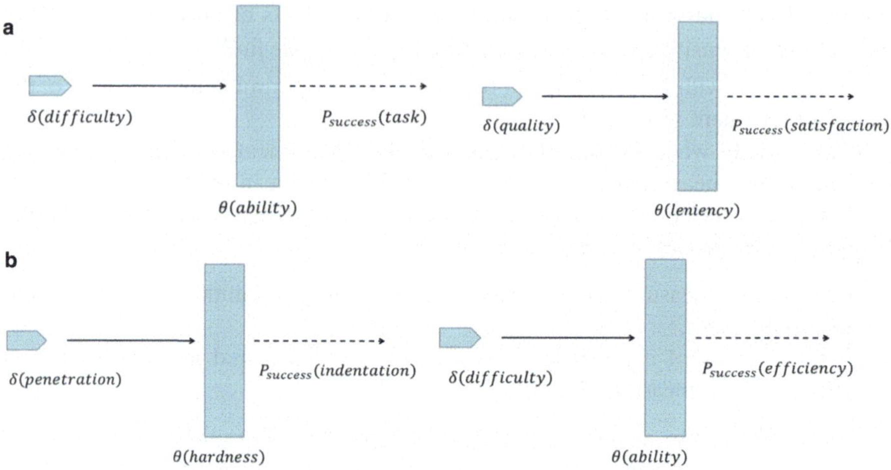

Fig. 4.11 Examples of qualitative item: probe systems

in Fig. 4.11, apart from human-based cases addressing task performance and satisfaction, the Rasch approach can also apply to other qualitative situations such as testing of material hardness blocks or the efficiency of an organisation, such as a hospital.

As in measurement in the physical sciences, it is a pre-requisite in any of these applications to achieve a clear separation between the measurement object and the instrument being used to probe the object, if one is to have a chance of establishing metrological references for traceability and comparability, as discussed further in the case study in Sect. 3.6.2.

4.5 Case Studies: Examples of Measurements

4.5.1 Physical Measurements

The measurement set-up for the example of weighing pre-packaged goods to check whether the packets contain the required mass is stipulated in Table 2.5, as exemplified in E4.1.

E4.1.1 Measurement system analysis: physical measurements

Choose any measurement situation: It can be measurements of the product you have chosen.	Your answers ... Coffee powder pre-packaged. Each packet should contain as promised: correct amount powder - 500 g
Make a summarising measurement system analysis:	
• Identify the principal elements of the measurement system (object, instrument, operator etc.)	Packet - Balance - Appraiser; environment, method

An overview of the three steps for implementation of measurement is given in Fig. 4.1: (A) calibration of a set of standards; (B) calibration and testing of a measurement system at the various levels; (C) implementation to measure the value of the measurand at each level of interest.

Calibration of a Set of Standards

Assuming that product consists of a variety of packaged quantities—say 1000 g, 500 g, 200 g, 100 g—then a first step in implementing processes for monitoring, measurement, analysis needed to demonstrate that the product conforms to requirements, is to calibrate a set of standards of mass to provide sufficient quality for the subsequent weighings of product against specification. The calibration process for the actual measurement system is of the kind marked 'Calibration' in Fig. 3.2. An instrument and a standard have to be used for the calibration. The instrument could be chosen to be the same as used subsequently in measurement of product, but often—for instance in an industrial shop-floor environment—calibration is made on another instrument, perhaps in a separate, quieter room (even in some cases at an independent calibration laboratory). The standard could in general be either an 'object', such as a weight; an instrument; or a reference procedure (such as a reference material, (Sect. 4.3.2)).

Good metrological practice is to match as far as possible the level of the standard—for mass in the present case—to the corresponding level of the product quality characteristic to be measured. This procedure minimises the 'distance' between the calibration and measurement levels which in turn minimises potential errors arising from not allowing for changes in the metrological characteristics of the measurement system which might occur as the level of measurement changes (Fig. 4.12).

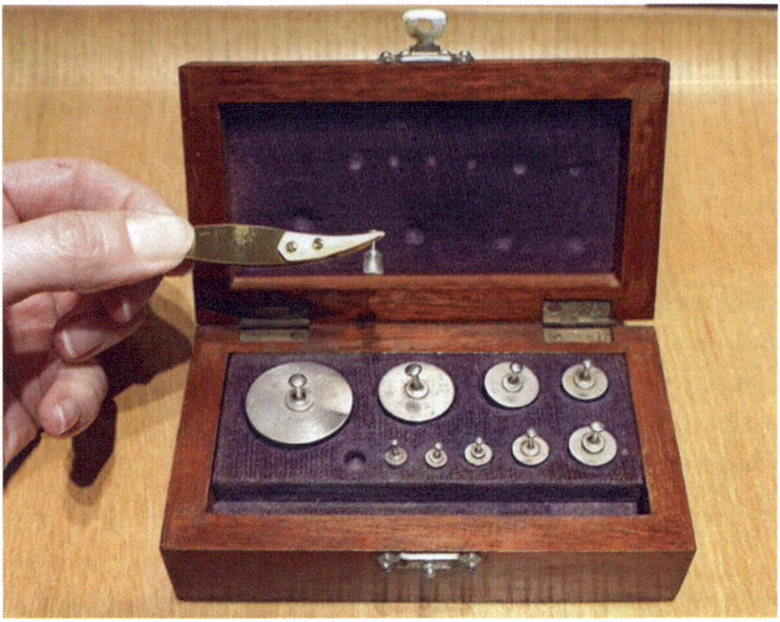

Fig. 4.12 Set of mass standards [ClipArt]

A common procedure in mass metrology is, starting from a 'head-weight', e.g. 1000 g, perform sub-division in order to calibrate a set of mass standards spanning the range of levels of interest. This ensures dissemination of metrological traceability described in Sect. 3.3. Sub-division consists of a series of different combinations of weighing the mass standards—say, in the range 100 g–1000 g—which is designed ideally to be 'complete'—i.e. sufficient to calibrate each mass standard—as well as minimising correlations between the different measured combinations. Such schemes are termed 'orthonormal' as described for example by a design matrix (Ivarsson et al. 1989) (Table 4.4): where each column indicates which mass standard is present (1 or −1) on either side of the weighing balance or is absent (0) in each sub-division ('deflection, *d*,') combination shown for each row of the design matrix. For example, deflection d1 is the observed difference in mass between on the one hand a 1 kg mass and on the other hand two 500 g weights; d8 weighs 200 g against another 200 g weight. An underlying assumption in making such a weighing design is that mass as a quantity can be added and subtracted (Rossi 2014 Sect. 1.2 *Counting and Measuring*).

The aim of the sub-division is to derive the correction, *c*, to be made to obtain the correct mass of any weight in the sub-division from the complete set of deflections, ***d***. The estimated corrections ***c*** to the each item attribute $\delta' = \delta - c$ are got from a least-squares regression of the corrections $\boldsymbol{c} = (\boldsymbol{A}^T \cdot \boldsymbol{A})^{-1} \cdot \boldsymbol{A}^T \cdot \boldsymbol{d}$ which best fit the observed deflections, ***d***. A general expression for the covariance matrix ***V*(*c*),** on which uncertainties of each estimated correction ***c*** from this fit can be estimated, is given by:

Table 4.4 Orthonormal design matrix (Ivarsson et al. 1989)

A =	1 kg	500 g	500 g′	200 g	200 g′	100 g	100 g′
d0:	1	0	0	0	0	0	0
d1:	1	−1	−1	0	0	0	0
d2:	0	1	−1	0	0	0	0
d3:	0	1	0	−1	−1	−1	0
d4:	0	0	0	1	−1	0	0
d5:	0	0	0	1	0	−1	−1
d6:	0	0	0	0	0	1	−1
d7:	0	1	0	−1	−1	0	−1
d8:	0	0	0	1	−1	0	0
d9:	0	0	0	0	1	−1	−1
d10:	0	0	0	0	0	1	−1

$$\boldsymbol{V}(\boldsymbol{c}) = \left\langle (\boldsymbol{c}' - \boldsymbol{c}) \cdot (\boldsymbol{c}' - \boldsymbol{c})^T \right\rangle = \left(\boldsymbol{A}^T \cdot \boldsymbol{A}\right)^{-1} \cdot \boldsymbol{A}^T \cdot \boldsymbol{V}(\boldsymbol{d}) \cdot \boldsymbol{A} \cdot \left(\boldsymbol{A}^T \cdot \boldsymbol{A}\right)^{-1}$$

which simplifies when all weighings are made on one instrument with a common variance $\boldsymbol{V}(\boldsymbol{d}) = \sigma^2 \cdot \boldsymbol{I}$ to the expression:

$$\boldsymbol{V}(\boldsymbol{c}) = \sigma^2 \cdot \left(\boldsymbol{A}^T \cdot \boldsymbol{A}\right)^{-1}$$

Orthogonality of the design matrix is ensured by duplicating a couple of weighing combinations: $d4 = d8$ and $d6 = d10$. With this ‘optimal’ scheme, all weights can be seen to participate in an equal number (five) of weighings. It is also straightforward to verify that the above design matrix is orthogonal (i.e. the vector products of the column vectors are zero), since $(\boldsymbol{A}^T \cdot \boldsymbol{A})^{-1}$ becomes a diagonal matrix, and no covariance is introduced from the least-squares regression.

A least-squares regression which allows for heteroscedasticity (i.e. different uncertainties) for the different weighing combinations yields estimates of the corrections, $\boldsymbol{c}$, for the N mass standards from observed deflections, $\boldsymbol{d}$, consisting of a design, $\boldsymbol{A}$, with P different weighing combinations:

$$\boldsymbol{c} = \left(\boldsymbol{A}^T \cdot \boldsymbol{W} \cdot \boldsymbol{A}\right)^{-1} \cdot \boldsymbol{A}^T \cdot \boldsymbol{W} \cdot \boldsymbol{d}$$

Weighting of the least-squares solution is done in terms of the observed standard deviations, s, in each weighing combination, j:

$$W_{j,j} = \left(\frac{1}{s_j}\right)^2 \cdot \left[\overrightarrow{\sum s \cdot s} - \langle s_0 \rangle^2\right]; j = 1 \ldots P$$

The correction, c_O, of the head-weight, which acts as a restraint for the solution, is added according to Gauss-Markov, to the vector of deflections as d_O, which is given infinite weight.

Although weighting to a certain extent breaks the orthogonality of the design matrix, it is still felt to be best practice to employ an initial orthonormal design since this should minimise covariances (Ivarsson et al. 1989).

Calibration and Testing of a Measurement System

If we were to rely on the instrument manufacturer's specifications, then typical values for the characteristics of the weighing machine used are known to be: *resolution 0,2 g; sensitivity, $K_{cal} = 1$ [Table 2.4].* . Restitution of the value of the measurand, z from the instrument display, y, is then straightforward using the expression (2.1): $z = S = K_{\mathrm{cal}}^{-1} \cdot R = K_{\mathrm{cal}}^{-1} \cdot y$ and assuming there is no bias to be corrected for.

Following the calibration of a set of mass standards [(A) above], the actual measurement system can be calibrated more carefully at each level, j, of interest; that is, 1000 g, 500 g, 200 g, 100 g. Restitution of the value of the measurand, z, from the instrument display, y, is then straightforward at each weighing level, j, using the expression (Sect. 2.4.4):

$$z_j = S_j = \left(\frac{R - b}{K}\right)_j = \frac{y_j - b_j}{K_j} \tag{2.9}$$

In general both the system sensitivity K_j and the bias b_j will in principle be different for each weighing level, j.

Implementation to Measure

Typically following calibration of the weighing machine (B), a series of n repeated measurements, i, of each entity of interest (packets of coffee in the present case) will be performed. It will often be observed that each new reading, y_i, will differ from previous readings (Table 4.5). Unless one expects a dynamic result, this scatter may either be a sign of changes in the value of the quality characteristic of the measured object or apparent changes arising from limited measurement quality, that is, measurement uncertainty, as already discussed in Chap. 2. The measurement readings given in Table 4.5 are those following restitution using Eq. (4.1) at the 500 g-level.

Table 4.5 Example of repeated measurements (weighing of pre-packaged goods)

Observation	Measurement reading, z_i (mass in g)	
1	495	
2	497	
3	493	
4	496	
Mean	495.3 $n = 4$	$\bar{z} = \frac{1}{n} \cdot \sum_{i=1}^{n} z_i$
Standard deviation	1.7	$s = \sqrt{\frac{1}{n-1} \cdot \sum_{i=1}^{n} (\bar{z} - z_i)^2}$
Standard deviation of mean	0.85	$s(\bar{z}) = \frac{s}{\sqrt{n}}$

Apart from registering a list of values (as in Table 4.5), it is also useful to make a plot showing graphically how the repeated values are distributed. This can be done in the form of a histogram where the height of each point on the vertical axis shows the number of times that particular value (on the horizontal axis) is obtained. In cases where a limited number of observations are made, the frequency with which the value is obtained can be replaced by an estimated probability (in which case the curve displays the probability distribution function (PDF) (Sect. 4.2.1).

The ultimate aim of the measurements in this example of pre-packaged goods is to verify that the contents lie within legal specifications, that is, the mass of each packet should exceed a lower specification limit $L_{SL,\bar{x}}$ (shown in Fig. 4.13). This mass for each packet will be the mean (or average) $\bar{z}$ given in Table 4.5. A measure of the quality of this determined value will thus be the standard deviation in the mean, $s(\bar{z})$, as also calculated in Table 4.5. Additional contributions will arise from uncertainties Eq. (4.3) associated with each term in the restitution Eq. (4.1) and of course the calibration uncertainties from steps (A) and (B) above.

In the present case of mass measurements, Eq. (4.2) can be evaluated as follows to yield the relative standard uncertainty in the item attribute value of the mass of each coffee packet from a restitution process: $u(z = S) = u\left(\frac{R-b}{K_{\mathrm{cal}}}\right) = \frac{R}{K_{\mathrm{cal}}} \cdot \sqrt{\frac{u(R-b)^2}{(R-b)^2} + \frac{u(K_{\mathrm{cal}})^2}{K_{\mathrm{cal}}{}^2}} = \frac{R}{K_{\mathrm{cal}}} \cdot \frac{u(R-b)}{R-b} = \frac{500\ \mathrm{g}}{1} \cdot \frac{1.3\ \mathrm{g}}{500\ \mathrm{g}-5\ \mathrm{g}} = 13\ \mathrm{g}$,

where $u(R-b) = \sqrt{u(R)^2 + u(b)^2} = 1.3$ g ; $u(R) = 0.85$ g from the standard deviation of the mean mass (Table 4.5), at a response level $R = 500$ g of the weighing machine; $u(b) = 1$ g is the uncertainty on the calibrated bias, b, (typically 5 g) of the weighing machine; and $u(K_{\mathrm{cal}})$~0. It is assumed in this case that $(R - b)$ and K_{cal} are uncorrelated.

Further discussion of the evaluation of measurement uncertainty in this example will be given in Chap. 5, to be followed in Chap. 6 by a final discussion of the risks of incorrect decisions of conformity.

$$g_{entity}(x) = \frac{1}{\sqrt{2\pi}\cdot s_p} e^{-\frac{(x-\hat{x})^2}{2\cdot s_p^2}}$$

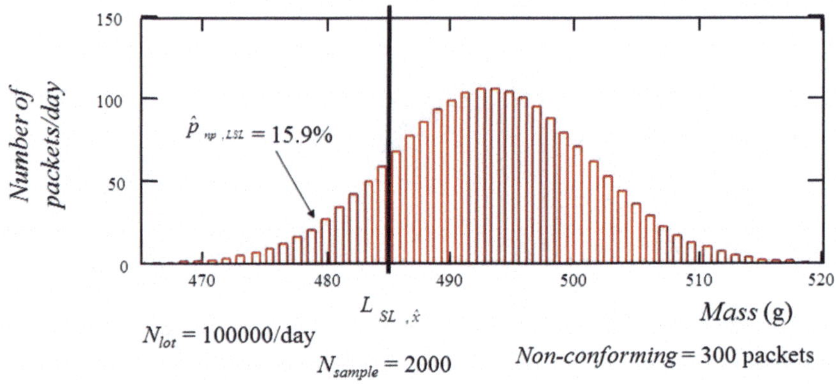

Fig. 4.13 Example of distribution of repeated measurements (weighing of pre-packaged goods)

4.5.2 Human Challenges

Basically the same three steps (Fig. 4.1) are also applicable for describing the implementation of processes to measure quantities associated with human challenges. How to treat measurement where a human acts as a measurement instrument will be illustrated for the example of pre-packaged goods.

E4.1.2 Measurement system analysis: measurements with humans

Choose any measurement situation: It can be measurements of the product you have chosen.	Your answers …*Coffee powder pre-packaged*. One experiences a certain 'prestige' when serving this mark of coffee: Measure of perceived 'prestige': 80%[1], at least 60% (customer survey)
Make a summarising measurement system analysis:	
• Identify the principal elements of the measurement system (object, instrument, operator etc.)	***Measurement object***: *samples of pre-packaged coffee packets; quality characteristics: quality associated with prestige* ***Measurement instrument***: *Panel of consumers; quality characteristics: leniency*[2]. *Perceived prestige response of each panellist to packet quality registered with a questionnaire with 10 items, and each item graded with 3 categories (0,1,2). Decision-making accuracy in terms of 'correctness' of categorisation,* $P_{success}$*, from* $\theta - \delta = \log\left(\frac{P_{success}}{1-P_{success}}\right)$ Eq. (1.1) or Eq. (2.11) (Sects. 2.4.4 and 2.4.5)

[1] On the scale of 'prestige' which is rated 0 –100%
[2] 'Leniency' = how easily satisfied

(A) Calibration of a Set of Standards, Prestige

Assuming that product consists of a variety of packaging with a range of prestige, then a first step in implementing processes for monitoring, measurement, analysis needed to demonstrate that the product conforms to requirements, is to calibrate a set of standards for prestige to provide sufficient quality for the subsequent measurement of product prestige against specification.

The calibration process for the actual measurement system is of the kind marked 'Calibration' in Fig. 3.2. Obviously the calibration process requires, as explained in Chap. 3, the capability to make separate assessments of the characteristics of the instrument used (in the present case, person leniency to prestige) and those of the measurement object (property of the product to stimulate perceived prestige).

Examples of earlier studies of consumer preference concerning packaging include the work of Camargo and Henson (2015) and Tarka (2013). A first step is to formulate a construct specification equation Eq. (4.6) for the construct 'prestige', z, as a function of a number of explanatory variables, $\boldsymbol{x}$ and $\boldsymbol{y}$. A Table 4.6 can be set up corresponding to the first row of Table 1.2 for customer satisfaction.

Adapting the prestige questionnaires to the current case of prepacked goods, one could formulate a 10-item questionnaire:

1. My attention was drawn to this particular brand
2. I already know how the product looks like
3. I like the look of the product in this packaging
4. Holding the packaging in my hand gives a prestigious sensation
5. The colour of the packaging has a prestigious feel to it
6. The product in this packaging is likely to produce coffee which tastes and smells delightful
7. The packaging of the product makes me feel like I would be buying a great product
8. The product in this packaging could give me a refreshing sensation
9. If I had the money, I would buy this product
10. The font of the text on the packaging has a prestigious feel to it

Factors which might determine the perceived prestige of a packet include choice of colour or nuance; surface quality (e.g. glossy or smooth paper); type font, etc. Each of the 10 items chosen in the questionnaire in this example has been formulated to address a certain level of prestige, as determined by some combination of contributions from each explanatory variable for both item, δ, and person (θ, probe) (Table 4.6). Further evaluation of the construct specification equations, in which the various coefficients for the different explanatory variables are evaluated with principal component regression, will be described in Chap. 5.

Table 4.6 Item and person attributes explaining perception of prestige

Construct: prestige	Questionnaire item	Item characteristic, $\delta = f[x_1, \ldots, x_m]$	Person characteristic, $\theta = g[y_1, \ldots, y_m]$
General attention to attractiveness	1, 2, 6, 7, 8	x_1: Ease of item to attract attention	y_1: Sensitivity of person to product attractiveness
Visual perception	3, 5, 10	x_2: Visual attractiveness of product	y_2: Sensitivity of visual perception of person
Tactile perception	4	x_3: Tactile attractiveness of product	y_3: Sensitivity of tactile perception of person
Uncontrolled variable	9	x_4: Cost of production	y_4: Wealth of person

Once again, as in the sub-division case above, it is good metrological practice to match as far as possible the level of the standard—for prestige in the present case—to the corresponding level of the product quality characteristics to be measured. To make the construct specification as complete and representative as possible, one is recommended to follow the procedures outline in Sect. 5.2.1, 'Construct description. Prioritisation'.

If this is the first time traceable measurements are attempted for 'prestige', the normal procedure in psychometrics is to calibrate a set of standards by making a matrix of measurements where each instrument (person) is used to measure each standard at the various levels. In other words, steps (B) and (C) are performed at the same time as step (A).

(B) Calibration and Testing of a Measurement System

A logistic regression is performed simultaneously of the Rasch model Eq. (1.1) to the matrix of responses of each instrument (person, i) to the complete set of measurement objects (items, j).

As described in Chap. 3, restitution will yield estimates, δ, of the respective properties of interest—in the present example, the capability of a packet to cause a certain level of perceived prestige (Fig. 4.14)—by applying the Rasch model (Eq. 1.1) to the response data exemplified in Table 4.7. or more properly the polytomous version Eq. (4.4) to account for the three categories—0,1,2 (disagree, neutral, agree)—of the response data, $q_{i,\, j,\, c}$.

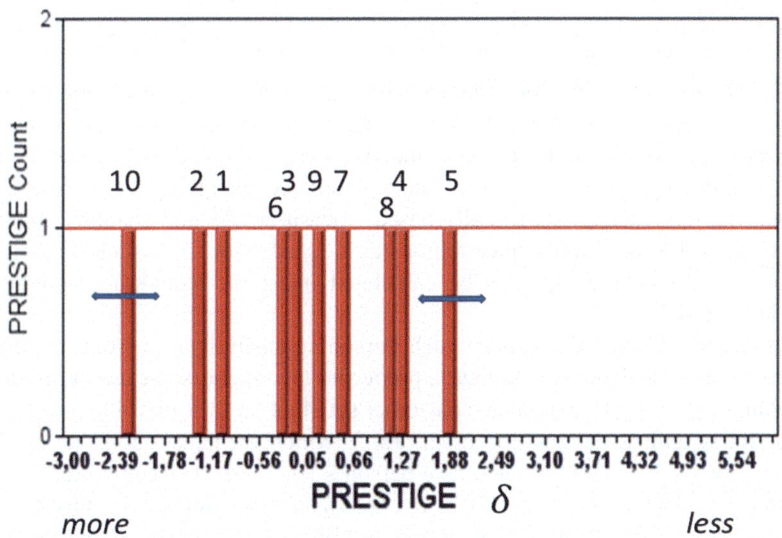

Fig. 4.14 Example of distribution of item attribute values for stimulating perceived prestige of pre-packaged goods for the set of 10 items, from raw data in Table 4.7 ($k = 2$, $N_{TP} = 73$)

Table 4.7 Example of repeated measurements (prestige of pre-packaged goods)—extract

Scores (0–2) 10 items	TP
1,2,1,1,1,0,2,0,1,2	A
2,2,2,2,2,2,2,2,2,2	B
2,2,1,1,0,1,1,0,1,2	C
1,0,1,0,0,1,0,1,2,2	D
1,0,1,0,1,0,1,0,0,1	E
1,0,1,1,2,1,1,0,1,1	F
2,2,2,0,0,2,2,2,0,2	G

The distribution of item characteristics shown in Fig. 4.14 indicates that item 10—effect of font on perceived prestige—has the highest level of attractiveness, while item 5—the colouring of product—has the least effect on attractiveness. Measurement uncertainties, indicated with double-ended arrows (coverage factor, $k = 2$) have been evaluated using Eq. (4.2). Such observations can of course be used to re-design the product in the future to enhance customer satisfaction.

(C) Implementation to Measure

In most Rasch analyses of this kind, it is common to assume that the ability, θ, is a characteristic of each person (Fig. 4.6) but does not depend on which item is being responded to. That is the specific objectivity assumption made by Rasch (1961). The accuracy, however, with which this calibration is performed will depend on the ranges: both the span of object attribute value and the span of person abilities.

Further discussion of the validity and reliability when applying Rasch restitution will be given in Chap. 5.

The distribution of person characteristics shown in Fig. 4.15 indicates the spread of leniency over the cohort of test persons on the same scale as item attributes (Fig. 4.14). Measurement uncertainties, indicated with double-ended arrows (coverage factor, $k = 2$) have been evaluated using Eq. (4.2). Apart from what can be considered 'statistically-based' measurement uncertainties shown in Figs. 4.5 and 4.6, additional uncertainties associated with for instance instrument (person) sensitivity and bias can be calculated using the techniques and models given in Sect. 4.4.4.

Understanding what determines each person's sensitivity to visual, tactile and other psychophysical and psychometric properties can of course be used to re-design the product in the future to enhance customer satisfaction. It is possible to formulate a construct specification equation (Sect. 4.4.3) which explains—both 'controllable and' 'uncontrollable' factors (such as a person's age, gender, wealth, etc.)—which can affect product perception. Chap. 5 contains further details for this example how a construct specification equation is established for 'prestige' and what the measurement uncertainties are. We therefore defer completion of the example

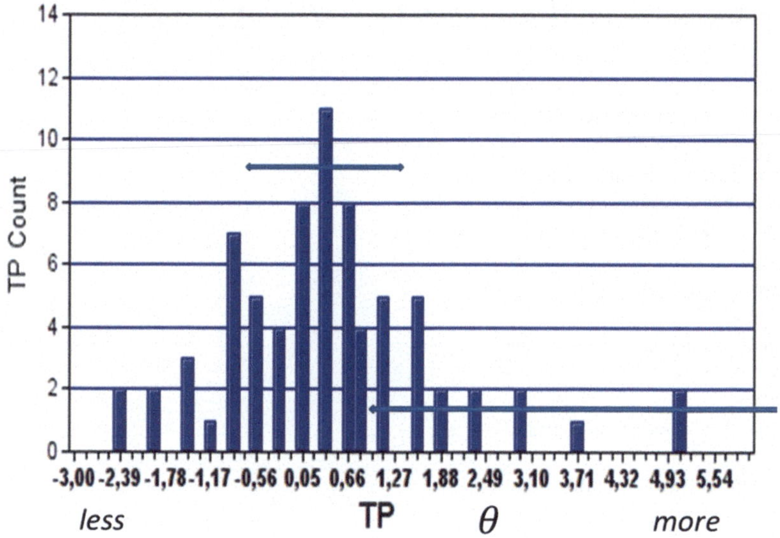

Fig. 4.15 Example of distribution of test person (TP) attribute values for person leniency to prestige of pre-packaged goods for the set of 10 items, from raw data in Table 4.7 ($k = 2$, $N_{TP} = 73$)

Table E4.2 *Expression of measurement uncertainty* for this second case study about *Human challenges* until that detailed analysis has been performed.

Exercises 4: Presentation of Measurement Results

E4.1 Measurement System Analysis

Choose any measurement situation: It can be measurements of the product you have chosen	Your answers..
Make a summarising measurement system analysis:	
• Identify the principal elements of the measurement system (object, instrument, operator, etc.)	
• Draw an Ishikawa diagram of the measurement system	
• Establish a measurement error budget	
Others:	

E4.2 Expression of Measurement Uncertainty

With your measurement data (possibly simulated) got from the measurement situation you chose in §4.1:	Your answers..
• Calculate the mean and standard deviation in each case you have repeated measurement values for a quantity	
• Correct each mean for known measurement errors	
• Express a standard measurement uncertainty for each source of measurement error:	
– Type-A evaluation for repeated measurements	
– Type-B evaluation in other cases	
– Combine the various standard measurement uncertainties	
– Calculate an expanded measurement uncertainty. Quote the coverage factor you have chosen	
Others:	

References

D. Andrich, A rating formulation for ordered response categories. Psychometrika **43**, 561–573 (1978)

E. Bashkansky, S. Dror, R. Ravid, P. Grabov, Effectiveness of a product quality classifier. Qual. Eng. **19**(3), 235–244 (2007)

BCR, Improvements and harmonization in applied metrology. in *EUR 9922,* Community bureau of reference, DG science, research & development. Commission of the European Communities, ed. by H. Marchandise (1985)

J.P. Bentley, *Principles of Measurement Systems*, 4th edn. (Pearson\Prentice-Hall, London\Upper Saddle River, 2004). ISBN-13: 978-0130430281, ISBN-10: 0130430285

L. Brillouin, Science and information theory, in *Physics Today*, vol. 15, 2nd edn., (Academic Press, Melville, 1962). https://doi.org/10.1063/1.3057866

F.R. Camargo, B. Henson, Beyond usability: Designing for consumers' product experience using the Rasch model. J. Eng. Des. **26**, 121–139 (2015). https://doi.org/10.1080/09544828.2015.1034254

S.J. Cano, J. Melin, L.R. Pendrill and The EMPIR NeuroMET 15HLT04 Consortium, Towards patient-centred cognition metrics, in *Joint IMEKO TC1-TC7-TC13-TC18 Symposium: "The future glimmers long before it comes to be"*, St. Petersburg, Russia, 2–5 July 2019 (2019)

P.H. Charlet, A. Marschal, Benefits of the implementation of a metrological structure for water analyses. Accreditation and Quality Assurance – Journal for Quality, Reliability and Comparability in Chemical Measurement **8**, 467–474 (2003)

E.N. Dzhafarov, Mathematical foundations of universal Fechnerian scaling, in *Theory and Methods of Measurements with Persons*, ed. by B. Berglund, G. B. Rossi, J. Townsend, L. R. Pendrill, (Psychology Press, Taylor & Francis, Milton Park, 2011)

EA-4/16, *Guideline on the Expression of Uncertainty in Quantitative Testing* (EA, Organisation for European Accreditation, Berlin, 2003)

EU commission, *Directive 2014/32/EU of the European Parliament and of the Council of 26 February 2014 on the harmonisation of the laws of the Member States relating to the making available on the market of measuring instruments* (2014)

EURACHEM/CITAC, *Traceability in Chemical Measurement—A Guide to Achieving Comparable Results in Chemical Measurement* (Eurachem/CITAC, Lisbon, 2003)

M. Golze, Why do we need traceability and uncertainty evaluation of measurement and test results? ACQUAL **8**, 539–540 (2003)

G. Gooday, The values of precision, ed. M. Norton Wise, Princeton University Press (1995) ISBN 0-691-03759-0

J.P. Guilford, *Psychometric Methods* (McGraw-Hill, Inc, New York, 1936), pp. 1–19

D.J. Hand, *Measurement – A Very Short Introduction* (Oxford University Press, New York, 2016)., ISBN 978-0-19-877956-8

J. Hannig, C.M. Wang, H.K. Iyer, Uncertainty calculation for the ratio of dependent measurements. Metrologia **40**, 177–183 (2003)

S.M. Humphry, The role of the unit in physics and psychometrics. Meas. Interdiscip. Res. Perspect. **9**(1), 1–24 (2011)

ISO 10012, *Measurement Management Systems – Requirements for Measurement Processes and Measuring Equipment* (International Standardisation Organisation, Geneva, 2003)

ISO 5725, *Accuracy (Trueness and Precision) of Measurement Methods and Results — Part 1: General Principles and Definitions, ISO 5725-1:1994(en)* (International Standardisation Organisation, Geneva, 1994). https://www.iso.org/obp/ui/#iso:std:iso:5725:-1:ed-1:v1:en:sec:A

ISO/IEC 17043, *Conformity Assessment - General Requirements for Proficiency Testing* (International Standardisation Organisation, Geneva, 2010)

S. Ivarsson, B. Johansson, H. Källgren, L.R. Pendrill, *Calibration of Submultiples of the Kilogram* (SP Report 1989:32 National Testing Institute, Borås, 1989)

G. Iverson, R.D. Luce, “The representational measurement approach to psychophysical and judgmental problems”, chapter 1, in *Measurement, Judgment, and Decision Making*, (Academic Press, Cambridge, 1998)

JCGM 100, *Evaluation of Measurement Data – Guide to the Expression of Uncertainty in Measurement (GUM 1995 with Minor Corrections)* (Joint Committee on Guides in Metrology (JCGM), Sèvres, 2008)

H. Källgren, M. Lauwaars, B. Magnusson, L. Pendrill, P. Taylor, Role of measurement uncertainty in conformity assessment in legal metrology and trade, Accreditation and Quality Assurance – Journal for Quality, Reliability and Comparability in Chemical Measurement, **8**, 541–7 (2003)

B. King, Metrology and analytical chemistry: Bridging the cultural gap. Metrologia **34**, 41–46 (1997)

B. King, Meeting ISO/IEC 17025 traceability requirements. Accred. Qual. Assur. **8**, 380–382 (2003)

K. Koffka, *Principles of Gestalt Psychology* (Harcourt, Brace, New York, 1935)

P.M. Kruyen, W.H.M. Emons, K. Sijtsma, Test length and decision quality in personnel selection: When is short too short? Int. J. Test. **12**, 321–344 (2012). https://doi.org/10.1080/15305058.2011.643517

S. Kullback, R. Leibler, On information and sufficiency. Ann. Math. Statist. **22**, 79–86 (1951). https://doi.org/10.1214/aoms/1177729694

J. M. Linacre, Bernoulli Trials, Fisher Information, Shannon Information and Rasch, Rasch Measurement Transactions **20**:3 1062–3 (2006), https://www.rasch.org/rmt/rmt203a.htm

J. Linacre, Sample size and item calibration stability. Rasch Meas. Trans. **7**(4), 328 (1994)

J. M. Linacre, W. P. Fisher, Jr, Harvey Goldstein’s objections to Rasch measurement: a response from Linacre, Fisher, Rasch Meas. Trans., **26**:3 1383–9 (2012)

G.N. Masters, A Rasch model for partial credit scoring. Psychometrika **47**, 149–174 (1982)

H.U. Mittmann, M. Golze, A. Schmidt, *Accreditation in global trade*, ILAC and IAF joint conference on 23–24 September 2002, Berlin, Germany. ACQUAL **8**, 315–316 (2003)

J. Moberg, C. Gooskens, J. Nerbonne, N. Vaillette, Conditional entropy measures intelligibility among related languages, in *Proceedings of the 17th Meeting of Computational Linguistics in the Netherlands*, (LOT, Utrecht, 2007). https://dspace.library.uu.nl/bitstream/handle/1874/296747/bookpart.pdf?sequence=2

D.C. Montgomery, *Introduction to Statistical Quality Control* (Wiley, Hoboken, 1996). ISBN: 0-471-30353-4

J. Nerbonne, W. Heering, P. Kleiweg, Edit distance and dialect proximity, in *Introduction to reissue edition, Time Warps, String Edits and Macromolecules: The Theory and Practice of Sequence Comparison*, ed. by D. Sankoff, J. Kruskal, (CSLI, Stanford, 1999).

L.R. Pendrill, Discrete ordinal & interval scaling and psychometrics, in *Métrologie 2013 Congress*, (CFM, Paris, 2013)

L. R. Pendrill, Meeting future needs for Metrological Traceability – A physicist's view, Accreditation and Quality Assurance – Journal for Quality, Reliability and Comparability in Chemical Measurement, **10**, 133–9, http://www.springerlink.com/content/0dn6x90cmr8hq3v4/?p=2338bc01ade44a208a2d8fb148ecd37aπ (2005)

L.R. Pendrill, Using measurement uncertainty in decision-making & conformity assessment, Metrologia, **51**: S206 (2014)

L.R. Pendrill, Assuring measurement quality in person-centred healthcare. Meas. Sci. Technol **29**(3), 034003 (2018). https://doi.org/10.1088/1361-6501/aa9cd2

L.R. Pendrill, W.P. Fisher Jr., Counting and quantification: Comparing psychometric and metrological perspectives on visual perceptions of number. Measurement **71**, 46–55 (2015)

Z. Pizlo, Symmetry provides a Turing-type test for 3D vision, in *Mathematical Models of Perception and Cognition*, ed. by J. W. Houpt, L. M. Blabla, vol. 1, (Routledge, Abingdon, 2016). ISBN 978-1-315-64727-2

R. Pradel, T. Steiger, H. Klich, Availability of reference materials: COMAR the database for certified reference materials. Accred. Qual. Assur. **8**, 317–318 (2003)

G. Rasch, On general laws and the meaning of measurement in psychology, 321–334 in Proceedings of the Fourth Berkeley Symposium on Mathematical Statistics and Probability, IV. Berkeley: University of California Press. Available free from Project Euclid (1961)

J.F. Reynolds, Estimating the standard deviation of a normal distribution. Math. Gaz. **71**, 60–62 (1987). https://doi.org/10.2307/3616296. https://www.jstor.org/stable/3616296

G. B. Rossi, Measurement and Probability – A Probabilistic Theory of Measurement with Applications, Springer Series in Measurement Science and Technology, Springer Dordrecht, https://doi.org/10.1007/978-94-017-8825-0 (2014)

K.-D. Sommer, M. Kochsiek, Role of measurement uncertainty in deciding conformance in legal metrology. OIML Bull. **XLIII**, 19–24 (2002)

K.-D. Sommer, B.R.L. Siebert, Systematic approach to the modelling of measurements for uncertainty evaluation. Metrologia **43**, S200–S210 (2006). https://doi.org/10.1088/0026.1394/43/4/S06

C. Spearman, The proof and measurement of association between two things. Am. J. Psychol. **15**, 72–101 (1904)

C. Spearman, Correlation calculated from faulty data. Br. J. Psychol. **3**, 271–295 (1910)

J. Stenner, M. Smith, Testing construct theories. Percept. Mot. Skills **55**, 415–426 (1982). https://doi.org/10.2466/pms.1982.55.2.415

P. Tarka, Construction of the measurement scale for consumer's attitudes in the frame of one-parametric Rasch model, in *Acta Universitatis Lodziensis Folia Œconomica*, vol. 286, (2013), pp. 333–340. http://dspace.uni.lodz.pl:8080/xmlui/handle/11089/10321?locale-attribute=en

L.L. Thurstone, The stimulus-response fallacy in psychology. Psychol. Rev. **30**, 354–369 (1923)

P. van der Helm, Simplicity versus likelihood in visual perception: From surprisals to precisals. Psychol. Bull. **126**, 770–800 (2000)

B.D. Wright, Comparing factor analysis and Rasch measurement. Rasch Meas. Trans. **8**(1), 3–24 (1994)

Chapter 5
Measurement Report and Presentation

Any report and presentation of measurement results will include both the measured value as well as some estimate of the measurement uncertainty in the measured value. In this penultimate chapter, and prior to the final step of decision-making covered in Chap. 6, we consider how best to model measurement information throughout the measurement process—from entity stimulus, through instrument response to restitution of the measurement value. A general and broad formulation is sought of differences applicable in both the physical and social sciences.

To provide measures of entity variation separate from apparent dispersion caused by limited measurement quality means making a clear distinction between:

- In the global scenario, where different entities[1] will have different quantity values—for instance, due to the effects of variation in production or wear and tear on a product.
- An imperfect measurement system on the other hand might change the classification result for the same property and entity from one category to another.

Measurement uncertainty, introduced in Chap. 4, is the latter which Helton (1997) classifies as subjective (i.e. epistemic) uncertainty (in contrast to the stochastic (i.e. aleatory) kind, see also van der Bles et al. (2019)).

Rossi (2014) in another book in this series gives an account of a probabilistic theory of measurement which will be recalled below. Rossi also emphasises the role of the notion of the measuring system with sufficient generality so as to apply also to perceptual measurement. Formulation of the measurement process as a concatenation of observation and restitution is a fundamental step and Rossi (2014 p. 126) makes clear that when we say 'measurement value', we are referring—not to the indication of the response of the measurement system—but in short to the restituted value.

[1] A single entity which changes its property values during a measurement is considered as different entities.

L. Pendrill, *Quality Assured Measurement*, Springer Series in Measurement Science and Technology, https://doi.org/10.1007/978-3-030-28695-8_5

In this chapter, we build on Rossi's approach when presenting measurement for both the physical and social sciences, extending the probabilistic theory for treating measurement error and uncertainty by using concepts of entropy and symmetry. Deploying the entropy concept (already used in Chap. 3 to describe measurement units) has the advantage of being applicable arguably to all scales—even the most qualitative scales such as the nominal. Error and uncertainty estimates in qualitative measurements can be expressed, respectively, in terms of the distortion, fuzziness and lack of clarity of a wide range of characteristics—patterns, shapes, smells, emotions and so on—using basic measures of information content not only with probability theory, but even possibility, plausibility and so on (Klir and Folger 1988; Schneider et al. 1986; Possolo 2018). Informational entropy has of course its roots in the analogous concept of entropy in thermodynamics, first coined by Carnot (1803) who was concerned that 'In any machine the accelerations and shocks of the moving parts represent losses of *moment of activity*. . . . In other words, in any natural process there exists an inherent tendency towards the dissipation of useful energy'.

The concept of entropy can in fact be invoked analogously to describe 'dissipation of useful information'—to paraphrase Carnot—at each of the three main stages in the measurement process—from (A) object, through (B) measurement to (C) response—not only to make a descriptive presentation (as in probability theory of measurement) but also in an explanatory and predictive way. For instance, when describing the quality characteristic of the measured entity, a task will be easier if there is some degree of order, i.e. less entropy. Similarly, a poorly performing measurement system can be explained in terms of both distortion and loss of information measured in terms of increases in entropy (Weaver and Shannon 1963), i.e. disorder, such as of a pattern.

The entropy concept can help when presenting measurement results at every level in the quantity calculus hierarchy (Table 3.1) from mere labels for nominal syntax, through semantic and pragmatic measures, to a full expression in terms of the nature of the quantity measured and effectiveness in changing conduct. This approach is particularly useful for prioritising (Sect. 5.2.1) among a plethora of potentially important factors (such as identified in a critical incidence process). This discussion will also prepare the ground for the final chapter of book, when decisions about product (in the widest sense) based on measurement need to be made.

5.1 Qualitative Measurement, Probability Theory and Entropy

5.1.1 Differences in Entity, Response and Measured Values: Entropy and Histogram Distances

Anywhere in the measurement process, the state of the measurement system will in general be characterised by a distribution over a range of categories described with a

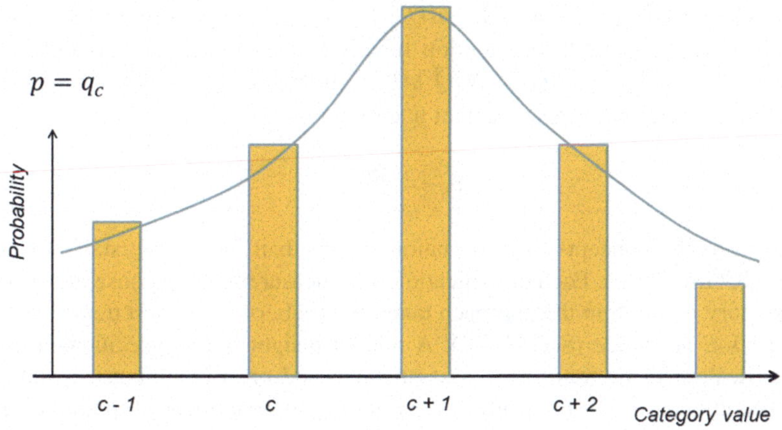

Fig. 5.1 Entropy (amount of information): $\Delta\mathbb{H}(Q) = -\sum_c q_c \cdot \ln(q_c)$ on a categorical scale

histogram (Fig. 5.1); a probability mass function (PMF) on a discrete, classification scale—which at the limit of a continuous scale becomes a probability density function (PDF). Two examples of such distributions can be found in Figs. 4.10 and 4.13. Distributions of this kind can in principle be found at every stage in the measurement process, from an a priori distribution of the entity attribute z to be measured through to the response, y, and final restitution, z_R, of the measurand and quality characteristic of the entity to be assessed (where the latter will be described in Chap. 6).

The amount of measurement information on the categorical scales of signals at any one point and state in the measurement process is in general the summed (change in) entropy, which for a discrete PMF is $\Delta\mathbb{H}(Q) = -\sum_c q_c \cdot \ln(q_c)$, where q_c is the occupancy of category c (Weaver and Shannon 1963). Information in each classification category, c, is expressed as a surprisal $-\ln(q_c)$, while the relative contribution to the total entropy is weighted with the relative occupancy, q_c.

In the limit where these discrete multinomial expressions invoked for categorical responses go towards the familiar continuous scales of traditional uncertainty presentations (Chap. 4):

$$\Delta\mathbb{H}(Q) = -\int_{-\infty}^{\infty} p(Q) \cdot \ln(Q) \cdot dQ = \ln\left[\sqrt{2\pi} \cdot u(Q)\right] + \frac{1}{2} \tag{5.1}$$

giving the integrated probabilistic (Shannon) formulation of the entropy across the width ('uncertainty') in the response. Equation (5.1) indicates that the two approaches—(1) standard uncertainty, u, and (2) decisions risks—can be unified.

In this chapter, examples of the different relations between entropy change and the natural logarithm of the width of a PMF (as depicted in Fig. 5.1) will be given for each measurement system element: entity (Sect. 5.2), instrument (Sect. 5.4) and rated response.

The relation $\Delta\mathbb{H}(Q) = \ln\left[\sqrt{2\pi}\cdot u(Q)\right]$ from Eq. (5.1) applies to the particular case where the probability distribution function of the outcome Q is taken to be Gaussian (Normal), i.e. $p(Q) = N\left[\overline{Q}, u(Q)\right]$. Inversion of Eq. (5.1) suggests an alternative expression of measurement uncertainty

$$u_q \sim e^{\Delta H(Q)} \tag{5.2}$$

more akin to the concepts of information theory than the classic standard uncertainties (JCGM GUM). Each classification of a measurement response into a particular category is found in the approach taken in this book to be best treated as either identification or choice (Sect. 6.3.2). A related insight is that specification limits, such as dealt with in conformity assessment based on a continuous, quantitative scale, become as 'marks on a ruler', thus uniting measurement of quantitative and qualitative properties. As pointed out already, the commonality between physical and social measurement (and qualitative estimations more generally) is first reached when one recognises that the *performance metrics* of a measurement system are the same concept in both (Pendrill 2014a, b). For this, we need to explicitly include *decision-making* as the third and final step—together with observation and restitution (Eq. 2.6).

We have a certain preference to express uncertainty in terms of an increase ΔH in entropy instead of a standard deviation (Zidek and van Eeden 2003 and Chap. 4) because it is conceptually closer to 'uncertainty' in everyday language—'decision quandary', is also substantially distribution-free and is indeed accessible to treatment not only with probability theory but also possibility and plausibility theories. Examples in which epistemic uncertainty is treated as synonymous with decision quandary can already be found in the literature: (Helton et al. 2006; Yang and Qiu 2005). This was mentioned in the Introduction to this book (Sect. 1.4) in connexion with product design considerations, which of course are analogous to measurement design.

At the same time, there is some reticence to use entropy in uncertainty considerations (Possolo and Elster 2014) and it is known that entropy-based distances, such as the Kullback–Leibler, are not true metrics, as will be discussed further in Sects. 5.5.1 and 5.5.3.

5.1.2 Differences in Measured Values at Each Stage of the Measurement Process

Typical measurement results derived from a Rasch analysis of (ordinal performance) test scores are illustrated in Fig. 5.2, where PMFs on common and linear scales show the distribution of the ability among the individual test persons (upper, blue columns) and the distribution of task difficulty (lower, red columns) for each item in two cases referred to in the present chapter: (a) the Knox Cube Test (KCT) and (b) the Functional Independence Measure (FIM) test. Another example—perceived prestige—was given at the end of Chap. 4.

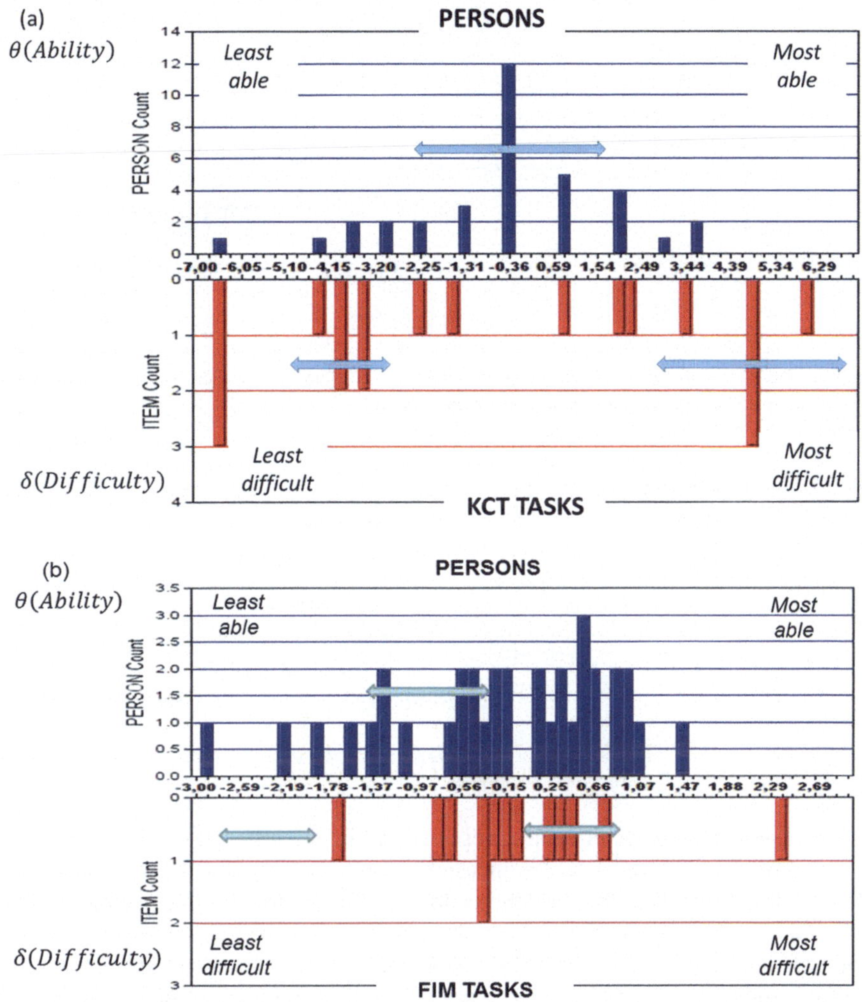

Fig. 5.2 PMF distributions of test person ability (upper, blue columns) and task difficulty (lower, red columns) in two cases: (**a**) the Knox Cube Test (KCT, Knox 1914) and (**b**) the Functional Independence Measure test (FIM, Linacre et al. 1994). Uncertainty coverage factor $k = 2$

Logistic regression consists of fitting Eq. (1.1) to the complete set of scores for every member of the cohort across the range of task difficulty spanned by the test items.

The extent to which the scale of scoring the response, $P_{success}$, for the various KCT tasks shows the effects of counted fraction non-linearity (Sect. 3.5.1) is illustrated in Fig. 5.3, in which the Rasch task difficulty (y-axis) is plotted against the classical test theory (CTT) mean, $\hat{x}_j = \frac{1}{N_{TP}} \cdot \sum_{i=1}^{N_{TP}} x_{i,j}$. [Data from WINSTEPS® example 'EXAM1.TXT'.]

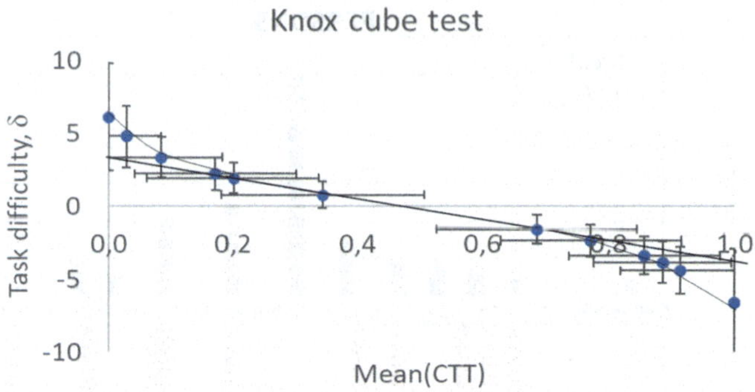

Fig. 5.3 Rasch task difficulty (y-axis) is plotted against the classical test theory (CTT) mean for the Knox cube test. Uncertainty coverage factors on both axes, $k = 2$

Fig. 5.4 Probabilistic and entropy models of the measurement system and processes (adapted from Rossi 2014, Fig. 5.5)

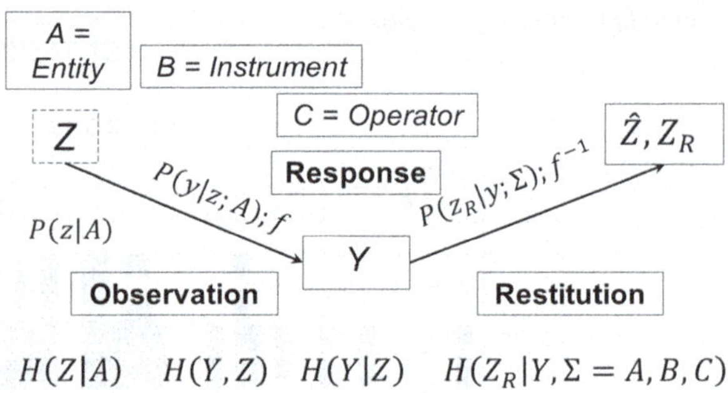

The typical measurement values presented in Fig. 5.2 reflect not only the task and person attributes intended to be measured, but also contain some aspects and limitations arising from imperfections in the measurement system employed to make the measurements. Throughout the passage of a measurement signal in our prototype measurement system (MSA and Fig. 2.4), the concept of entropy will be helpful in describing (and in some cases predicting) the amount of information at every stage, from the unknown stimulus from the entity to subsequent decision-making and restitution from the observed response. Ultimately the aim is to compensate as far as possible for the effects of imperfections in the measurement process, in order to obtain the most faithful measure of the quantities of interest: task difficulty and person ability in the two cases quoted.

A scheme of the measurement processes is given in Fig. 5.4, inspired in part by the probabilistic model of Rossi (2014; Fig. 5.5). The version shown here emphasises the role of each element of the measurement system, which will provide the steer for much of the material remaining to be presented in this book. From left to right in Fig. 5.4:

(a) With the quantity calculus case of an entitic quantity in mind, the quantity associated with an entity denoted A (Fleischmann's (1960) *Sachgrösse*; see

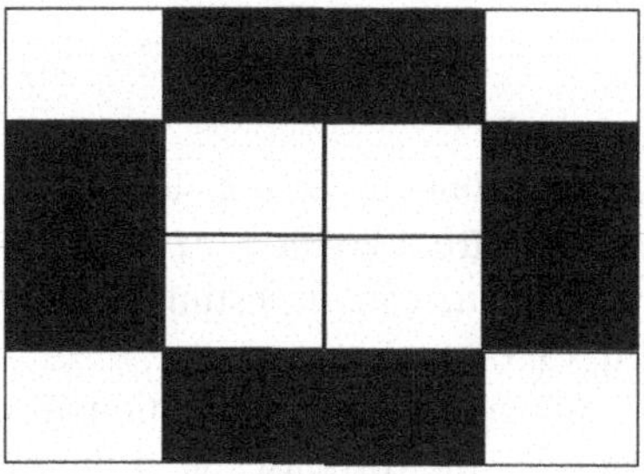

Fig. 5.5 Modelling a block sequence (adapted from Schnore and Partington 1967)

Chap. 3), alongside the (general) quantities—Z (stimulus) and Y (response)—of the measurement system (Sect. 5.2) exemplified with a case study (Sect. 5.3).

(b) The relation between the response and the stimulus is mainly determined by the characteristics of the instrument at the heart of the measurement system (Sect. 5.4).

(c) The third main element of the measurement system—the operator—plays an active role as rater (MSA 'appraiser') in judging the response and in performing final restitution of the measurand from the response (Chap. 6).

In Rossi's (2014) probabilistic theory of measurement, the link between the value, z, of the measurand, through the measurement system response, y, to the (restituted) measurement value, z_R, is described in terms of the joint probability distribution of discrete PMFs (Eq. (5.28) of Rossi (2014)):

$$P(z, y, z_R) = P(z) \cdot P(y|z) \cdot P(z_R|z, y) \tag{5.3}$$

Here P denotes a PMF where the height of the histogram column for a certain category is the *occupancy*, that is, the probability of obtaining that particular category.

This descriptive approach of probability theory is implemented when presenting a measurement result by assuming that an expectation measurement value can be meaningfully expressed Eq. (5.23), Rossi (2014) as:

$$\hat{z} = \mathbb{E}(z_R|y) \tag{5.4}$$

A simple example is the case of elementary counting, described in Sect. 4.4.1 where for each PMF illustrated in Fig. 4.10, the increasing difficulty experienced by Munducuru Indians in counting larger numbers of dots is evident from both the scatter and displacement of each PMF. One can calculate initially an expectation response value from the measurement system (in this case with a human being acting as a measurement instrument) for the number of dots in each measurement with the formula:

$$\mathbb{E}(y|z_M) = \sum\nolimits_{k=1}^{K} \frac{P(y_k|z_M) \cdot y_k}{K} \tag{5.5}$$

where counts y_k are registered over a range of numbers, $k = 1, \ldots, K$ as categories and the 'true' count is M dots. The expected number of dots counted, that is, the measurand, can be estimated from the restituted value $\mathbb{E}(z_R|y)$ Eq. (5.4) using Eq. (5.3).

An underlying assumption in all this is that the classification results in each case—at the response, y; at the measurand, z or restituted, z_R; or anywhere in the process—lie on quantitative scales, where distances between values can be meaningfully assigned. This would certainly be a valid assumption on the discrete but quantitative scale shown in Fig. 4.13 depicting the distribution of mass of different packets of pre-packaged coffee. It would also apply in the elementary counting example shown in Fig. 4.10 showing the distribution of counts.

Many measures of inter- and intrahistogram distances between PMFs (Sect. 5.5.1), when attempting to describe error and measurement uncertainty, respectively, are not fully applicable on the ordinal and nominal scales of categorical measurements where distances between categories are generally not known (as illustrated in Fig. 5.3 for the Knox cube test). (On nominal scales there are no meaningful distances at all.) In such, performance metric cases and on the most qualitative responses on nominal scales, the response y expressed in terms of a scored probability of success, P_{success}, does not in general lie on a quantitative scale, and expressions such as Eqs. (5.4) and (5.5) do not work (Svensson 2001).

As already observed in the Munduruku counting case (Pendrill and Fisher 2015) and which is also valid for the Knox cube test, what is interesting arguably in most such examples is not the number of dots or blocks in themselves, which are already known, e.g. $z_M = M\ dots$, but rather measures of the *ability* of each counter to count and the level of *difficulty* of each counting task.

A special approach is therefore needed to describe correctly a general measurement process. We consider that entropy adds value to a probabilistic theory description of the measurement process. Step by step in the passage of information, again as illustrated in Fig. 5.4, through our prototype measurement system, the terms in the well-known conditional entropy expression:

$$H(Q|P) = H(P, Q) - H(P) \tag{3.3}$$

where $Q = P(y|z)$ and $P = P(z)$,are added together (where probabilities are multiplied in Eq. (5.3) (Rossi 2014)), and each entropy term can be exemplified, as done below. Expression (3.3) simply states how the amount of information transmitted by a measurement system is the initial 'deficit' in entropy coming from prior knowledge, $H(P) = H(Z|A)$ of the measurand plus losses and distortions $H(P, Q) = H(Y, Z)$ from imperfections in the measurement process which increase entropy. Entropy will be found to be a key concept when handling the quality assurance of data, be it on quantitative or qualitative scales, throughout the measurement process, from stimulation to restitution.

5.2 A: Entity Construct Description and Specification

We start with the quality characteristic, δ, of the entity, A, of interest (product quality; task difficulty, etc.) and how the concept of entropy can assist in describing, predicting and prioritising the entitic construct.

Often as a prelude to a series of measurement, the all-important definition of what is actually intended to be measured will be in most cases a defining moment, literally speaking (Sect. 1.4). Among the requirements of validity and reliability for measurements in both the physical and social sciences (Roach 2006), of particular interest at this stage is: *Validity*—Degree to which one succeeds to measure what we intend:

- *face validity*: test appears to measure as intended,
- *content validity*: degree to which all necessary items are included in test,
- *criterion validity*: comparison with a 'gold standard',
- *construct validity*: convergency and discrimination when varying object.

5.2.1 Prioritisation

In describing the entity subject to conformity assessment, perhaps when a manufacturer or service provider, in order to 'tailor-make' a product by balancing production and consumption requirements, needs to formulate relations between different characteristics (both functional and non-functional (Sect. 1.4.5)), it is usually necessary to consider many and initially unstructured characteristics of the entity (Table 1.1). Some criteria for prioritising among them are needed when bringing structure, be it manufacturing or 'production' anywhere in the social and physical sciences.

Firstly, there should be an ambition to capture unconditionally as many relevant product aspects as possible—no amount of sophisticated analyses subsequently will be able to compensate for a component missed at any early stage. The 'critical incidents' technique (CIT) is one often used method of capturing a broad spectrum of experiences of the user. A critical incident is an event that is particularly satisfying or dissatisfying, as typically identified using content analysis of stories or vignettes (Bitner et al. 1990; Gremler 2004), rather than quantitative methods in the data analysis.

Structuring, prioritising and choosing among the many entity characteristics the end-user might mention may be done in different ways.

The Activity Inventory tool (Massof et al. 2007) is one approach to structuring constructs, where a bank with items describing everyday activities is arranged within a hierarchical framework. At the top level of the hierarchy, activities are organised by Massof et al. (2007) according to the objective they serve (Daily Living, Social Interactions or Recreation). A selection among the (several hundred) identified items is made; a few items are classified as 'Goals', which describe what the person is intending to accomplish (e.g. prepare daily meals, manage personal finances, entertain guests). The remaining items in the bank, grouped under the Goals, are classified as 'Tasks'. Tasks in the vision-related studies of Massof et al. (2007) describe specific cognitive and motor activities that must be performed to accomplish their parent Goal (e.g. cut food, read recipes, measure ingredients, read bills, write checks, sign name). Further discussion of the results of that study will be given in Chap. 6.

Each stakeholder group (e.g. patients) can be asked to rate the relative importance of each task as an aid to ranking and prioritisation. Tools for this include: Importance-Performance Analysis (Dagman et al. 2013), where a simple experiment might be to ask a person to place task or situation cards on a table, where the position of each card is determined on the vertical axis in terms of perceived difficulty of performing that specific task and location of the card on the horizontal axis indicates how important the performance of the task is rated. The results of this simple investigation can be analysed further using logistic regression Eq. (1.1).

Massof and Bradley (2016) report the estimation of the utility (i.e. a variable representing value, benefit, satisfaction, happiness, etc.) assigned by each patient to a hypothetical risk-free intervention that would facilitate each of the identified important and difficult Goals in the Activity Inventory for that patient. Another technique is the Analytic Hierarchy Process, where the many decisions to be made are broken down pairwise when ranking and prioritising among choices (Dolan 2008).

5.2.2 Entity Attributes, Construct Specification and Entropy

Structural properties of most quality characteristics, such as the entity attribute, δ, in conformity assessment can be formulated, albeit without the fundamental structures of universal truths (such as found in the laws of Nature, Chap. 3). A classic example is the field of chemometrics, where Wold et al. (2001) state[2]: 'Examples in chemistry include relating:

- Y = properties of chemical samples to X = their chemical composition,
- Y = the quality and quantity of manufactured products to X = the conditions of the manufacturing process,
- Y = chemical properties, reactivity or biological activity of a set of molecules to X = their chemical structure (coded by means of many X-variables)'.

Several aspects of how entropy can be usefully deployed in describing (and in some cases to predict) the value of δ have already been covered in Sects. 3.2.2, 3.2.3 and 4.4.2. For instance, a task is expected to be easier if there is some degree of order, i.e. less entropy, $H(P)$, where $P = P(z)$ is the probability distribution associated with the entity characteristic. (The corresponding role played by entropy in describing and predicting person ability, e.g. memory, is mentioned when dealing with Man as a measurement instrument in Sect. 5.4.)

[2]'Y' in Wold et al. (2001) corresponds to 'Z' in our notation.

To investigate how the construct associated with the measured entity is specified including entropy, the multivariate approach of forming a specification equation, introduced in Sects. 1.4, 3.3.2 and 4.4.3, is followed. A couple of case studies illustrate typical considerations when forming construct specification equations: (1) the perceived prestige of packets of pre-packaged coffee (Sect. 4.5.2) and (2) the level of difficulty of remembering sequences, exemplified here with the classic Knox cube test (Fig. 5.2a).

Information theory draws analogies between thermodynamic entropy in physics and the measures of information content in communication systems (Weaver and Shannon 1963). Information content can range from basic examples, such as the number of elementary symbols, to increasingly sophisticated information, through syntax, semantic and pragmatic aspects of meaningful information in many contexts. Whether one is tasked with explaining the level of difficulty of remembering, say, a particular sequence (tapped blocks; series of numbers or words) or perhaps trying to understand what factors determine how attractive a particular product is perceived, the basic idea is that task 'difficulty' is proportional to the entropy, as follows.

The Shannon entropy is proportional to ln(P), where P is the probability of a message (Weaver and Shannon 1963). The less expected a message is (i.e. smaller P), the greater the amount of information conveyed ('surprisal'). Taking the logarithm facilitates addition and subtraction of different amounts of information.

Looking more closely at the concept of entropy and information content, Attneave (1954), Barlow (2001), Miller (1956) and others addressed how recognisable ('meaningful') patterns can improve communication by exploiting redundancy or 'chunking'. Consider in general a message in which a number, N_j ($j = 1,\ldots,M$) of symbols of M different types can be distributed in a number, G, of categories (or cells) $G = \sum_{j=1}^{M} N_j$. The probability of encountering the jth symbol is $p_j = \frac{N_j}{G}$, which can be summed to unity. The total number, P, of messages that can be obtained by distributing the symbols at random over the G cells (with never more than one symbol per cell) is $P = \frac{G!}{\prod_{j=1}^{M} N_j!}$ (Brillouin 1962). The information theoretical entropy, which is a measure of the amount of information in these messages, is then:

$$
\begin{aligned}
I = -K \cdot \ln P &= -K \cdot \left[\ln (G!) - \sum_{j=1}^{M} \ln (N_j!) \right] \\
&\cong -K \cdot \left[G \cdot \ln (G) - \sum_{j=1}^{M} N_j \cdot \ln (N_j) \right]
\end{aligned}
\tag{5.6}
$$

where K is an arbitrary constant. Stirling's approximation in the final terms of Eq. (5.6) applies when G and N are large, but with modern computer power the approximation is no longer necessary when evaluating the factorial terms.

The difficulty, δ, of remembering a particular sequence of taps on a set of blocks in the classic Knox cube test for cognitive memory function (Stenner and Smith 1982; Melin et al. 2019) is one case study. Specification equations for task difficulty of this kind can also be developed for many other sequence memory tests: Apart from shorter sequences of tapped blocks in the KCT (and the similar Corsi block test) being easier to memorise, words at the start and ends of lists of the Auditory Verbal Learning Test (AVLT) are known to be easier to remember for most people than words in the middle, reflecting the effects of primacy and recency. Such relations between sequence length and entropy are captured by the first term $-\ln(G!)$ on the RHS of Brillouin's expression Eq. (5.6).

The Digit Span Test (DST) in addition has some sequences which include the *same* symbol more than once. The second term on the RHS of Brillouin's (1962) expression (5.6) then comes into play by reducing the entropy (making the tests easier) because it is considered easier to remember a repeated symbol, as captured by the term $-\sum_{j=1}^{M}\ln\left(N_j!\right)$.

Thus, similar arguments to Brillouin's (1962) would capture the reduction in entropy associated with the easily recognisable message compared with the other two messages consisting of less structured information. Continuing this line of thought, then a measurement unit—a kind of 'word'—could be described as reducing the entropy in a measurement 'message'. These ideas can even be developed when modelling reading comprehension, where assimilation of the information contained in whole texts is described in terms of a finite size memorised by the reader of an input set of propositions ('*n*-tuples of word concepts, one of which serves as a predictor (such as verbs, adjectives or conjunctions) and the other as arguments (usually nouns)') (Kintsch 1974; Latimer 1982). Our proposal would be to express the entropy of each proposition in the text as a measure of the difficulty of comprehension.

To capture in full generality this concept of efficiently carrying an amount of information, recourse can be made to similarity transformations in matrix algebra (Sect. 3.6.1). An arbitrary representation $\boldsymbol{v}$ (or pattern or message) can in general be decomposed into subsets if a similar (or displacement) matrix $\boldsymbol{D}$ can be found for the similarity transformation $\boldsymbol{v}'^{(a)} = \boldsymbol{D}^{-1} \cdot \boldsymbol{v}(a) \cdot \boldsymbol{D}$ which diagonalises every matrix in the representation into the same pattern of diagonal blocks—each of the blocks is a representation of the group independent of each other. When no further decomposition becomes possible, the representation is said to be irreducible.[3] Examples may be found in the study of symmetry in molecular or crystalline formations (Schneider et al. 1986). Analogies can be drawn with decomposing a message into words (and other structures—syntax, semantic and pragmatic—Sect. 3.2.1) and a measurement result in a set of measurement units. The connexion between measurement units and irreducible representations in quantum mechanics is discussed in Sect. 3.6.1.

Similar modelling of the difficulty of remembering block sequences was performed by Schnore and Partington (1967) who wrote in a description of the pattern shown in Fig. 5.5 meant to represent a block sequence:

[3]https://en.wikipedia.org/wiki/Irreducible_representation.

'For Pattern A, the occurrence of a black or white cell was determined randomly for the four cells in the upper left quadrant, with the constraint that two of the cells had to be black. The remaining quadrants of the pattern were obtained by reflecting vertically and then horizontally the quadrant for which the nature of the cells was determined randomly. Thus, Pattern A was symmetrical vertically as well as horizontally with the axes of symmetry passing between the second and third column and row. Type A patterns may be said to contain 2.6 bits of information because only six distinct patterns can be constructed under the rules outlined above: $\frac{4!}{2! \cdot 2!} = 6; \ln_2(6) = 2.6$ bits Eq. (5.6). The three quadrants of Pattern A which were obtained by reflection may be considered to be redundant and not adding any further uncertainty'. Schnore and Partington (1967) found that the number of recall pattern errors among 214 university summer school students increased in proportion to the task entropy for a series of patterns with successfully less symmetry.

5.2.3 Formulation of Construct Specification Equations

A construct specification equation (CSE, mentioned in the 'product' description Sect. 1.4, Eq. (1.2), and in Sect. 3.3.2) is often formulated in terms of a linear combination of a set of explanatory variables ***X***:

$$\hat{Y} = \sum_k \beta_k \cdot X_k \tag{5.7}$$

such as when describing a Rasch construct ***Y*** (e.g. task difficulty, δ, or person ability, θ).

A five-step procedure is followed to formulate this construct specification equation:

1. Rasch estimate, δ_j, for each item, j, based on an initial logistic regression of Eq. (1.1) to the response data for P_{success}.
2. Identify a set of explanatory variables, X_k, to inform a model of the expected item attribute (Sect. 5.2.1).
3. Principal component analysis among the set of explanatory variables, X_k (Sect. 5.2.5).
4. Linear regression of the Rasch estimates, δ_j (from step 1) against X' in terms of the principal components P identified at step 3 and the original data, X: $X' = X \cdot P$.
5. Conversion back from principal components to the explanatory variables, X_k, in order to express the construct specification equation for the item attribute in the form: $\hat{Y} = \sum_k \beta_k \cdot x_k$.

The construct specification equation Eq. (5.7) 'sets forth a theory of item-score variation and simultaneously provides the vehicle for confirmation or falsification of the theory' (Stenner and Smith 1982, 1983). At the highest level of construct theory

(Stenner et al. 2013), three central benefits are obtained, be it a traditional measurement system for measuring temperature or a psychometric measurement system. There is: (1) an observational outcome, often a count; (2) a causal mechanism that transmits variation in the intended attribute (i.e. reading difficulty or temperature) of the measurement object, via an instrument of sensitivity K, to the observed outcome, R; and (3) an object attribute measure of stimulus S (e.g. task difficulty or temperature) denominated in some unit (e.g. Lexiles or degrees Celsius) which is estimated from restitution from an observed response, R, of a previous calibrated measurement system of known sensitivity, K_{cal}. The construct specification equation 'breaks the relationship between intention (e.g. task difficulty) and attainment (e.g. student ability) by independently formalizing intent as an equation that can produce item calibrations or ensemble means that are in close correspondence with empirical estimates' (Stenner et al. 2013).

5.2.4 *Rasch-Based Construct Specification Equation from Logistic Regression*

An early formulation of a Rasch-based construct specification equation was done by Scheiblechner (1971) where an attribute value, δ_j^*, for the jth item is expressed as a sum of a number of explanatory variables, X_k,

$$\delta_j^* = -\left(\sum\nolimits_k \beta_{j,k} \cdot x_k + \varepsilon_\delta\right) \tag{5.8}$$

Values of the coefficients, β, in Eq. (5.8) were found by regression, minimising the sum of weighted squared residuals $\delta_j - \delta_j^*$ for the item attribute with respect to the Rasch estimate, δ_j, for each item, j, based on an initial fit of Eq. (1.1) to the response data for $P_{success}$.

$$\sum_k I_j \cdot \left(\delta_j + \sum_k \beta_{j,k} \cdot x_k + \varepsilon_\delta\right)^2 = \sum\nolimits_k I_j \cdot \left(\delta_j - \delta_j^*\right)^2 = \text{minimum} \tag{5.9}$$

and I_j is a measure of the information in the sample of subjects tested (Fischer 1973).

5.2.5 *Principal Component Regression*

The initial set, $\boldsymbol{X}$, of explanatory variables in the CSE of Eq. (5.8) may however exhibit correlation, making it unsuitable for the direct regression described in Sect. 5.2.4. Principal component analysis (PCA), where a matrix $\boldsymbol{P}$ of the principal components of variation is formulated, can be used in that case to transform $\boldsymbol{X}$ into an orthonormal set $\boldsymbol{X}'$:

$$\boldsymbol{X}' = \boldsymbol{T} = \boldsymbol{X} \cdot \boldsymbol{P}$$

The principal components of variation are the eigenvectors, $\boldsymbol{p}$, of the covariance of $\boldsymbol{X}$, with eigenvalues λ:

$$Cov(\boldsymbol{X}) \cdot \boldsymbol{p}_n = \lambda_n \cdot \boldsymbol{p}_n$$

As a second step, the Rasch construct $\boldsymbol{Y}$ (from Eq. (1.1), e.g. task difficulty, δ, or person ability, θ, with ε variation) is expressed as:

$$\boldsymbol{Y} = \boldsymbol{T} \cdot \boldsymbol{C} + \varepsilon_y$$

by performing a least-squares regression against the principal components:

$$\hat{\boldsymbol{C}} = \left(\boldsymbol{T}^T \cdot \boldsymbol{T}\right)^{-1} \cdot \boldsymbol{T}^T \cdot Y \tag{5.10}$$

The final step in formulating a CSE is to transform back into the measurement space:

$$\hat{\boldsymbol{Y}_0} = \boldsymbol{X}_0 \cdot \boldsymbol{P} \cdot \hat{\boldsymbol{C}}$$

to yield the CSE as a linear combination of the explanatory variables, $\boldsymbol{X}$:

$$\hat{Y} = \sum_k \beta_k \cdot x_k \tag{5.7}$$

where the coefficients in the linear predictor (construct specification equation) $\beta = \boldsymbol{P} \cdot \hat{\boldsymbol{C}}$.

5.3 Case Study: Knox Cube Test and Entity Construct Entropy

A principal thesis when formulating construct specification equations is that the concept of entropy, as a measure of the amount of information and order, can explain a construct attribute associated with the measured entity, such as the difficulty of a series of tasks. The Knox cube test (KCT) offers a suitable test of modelling task difficulty when increasingly complex sequences of sequences of taps are to be recalled, in which the Brillouin (1962) expression (5.6) for entropy can be evaluated for a number of sequences which not only have increasing numbers of taps but also some repeated taps of the same block. In particular, the entropy term is $\delta = -\left[\ln(G!) - \sum_{j=1}^{M} \ln\left(N_j!\right)\right]$ which follows the general form of Eq. (5.1).

A KCT sequence, such as '14': 1-4-2-3-4-1, can be expected to be easier to recall according to Brillouin's expression than a sequence of the same length (six taps) but without repeats. The expected reduction in entropy for this sequence is 1.4 compared with a $G = 6$ sequence without repeats. A Rasch analysis (logistic regression of Eq. (1.1)) of the raw data yields for this task $\delta = -3.4(1.4)$ *logits*, uncertainty coverage factor, $k = 2$. The task difficulty, δ_j, of the tasks, j, are shown in Figs. 5.2a and 5.3.

When attempting to explain these task difficulties (Melin et al. 2019) with a set of three explanatory variables: $X = \{\text{Entropy}, \text{Reversals}, \text{Average Distance}\}$, as a result of step 3 in the formulation of a CSE (Sect. 5.2.3) it was found that, while there is relatively high degree of correlation (Pearson coefficient $R = 0.87$) between entropy and reversals, as shown by the correlation matrix $\text{corr}(X) = \begin{pmatrix} 1 & 0.871 & 0.475 \\ 0.871 & 1 & 0.583 \\ 0.475 & 0.583 & 1 \end{pmatrix}$, the corresponding degree of correlation between entropy and average distance was considerably less ($R = 0.475$). In a subsequent principal component analysis, by seeking the eigenvalues and eigenvectors of the covariance matrix cov *(X)*, the matrix P consists of the three principal components of variation as three column vectors:

$$P = \begin{pmatrix} 0.795 & -0.606 & 0.039 \\ 0.602 & 0.779 & -0.172 \\ 0.073 & 0.16 & 0.984 \end{pmatrix}$$

The next step (4) in the formulation of a CSE (Sect. 5.2.3) was performed by making a linear regression of the Rasch task difficulties of the series of KCT sequences against the orthonormal set $X' = X \cdot P$. As a final step, the KCT construct specification equation (Eq. 5.11) was found by transforming the regression results back into the observed explanatory variables using the transformation: $\hat{Y} = zR = \sum_k \beta_k \cdot x_k$ Eq. (5.7), where the coefficients in the linear predictor (construct specification equation) $\beta = \boldsymbol{P} \cdot \hat{\boldsymbol{C}}$. The predicted values of KCT task difficulty are shown graphically in Figs. 5.6 and 5.9.

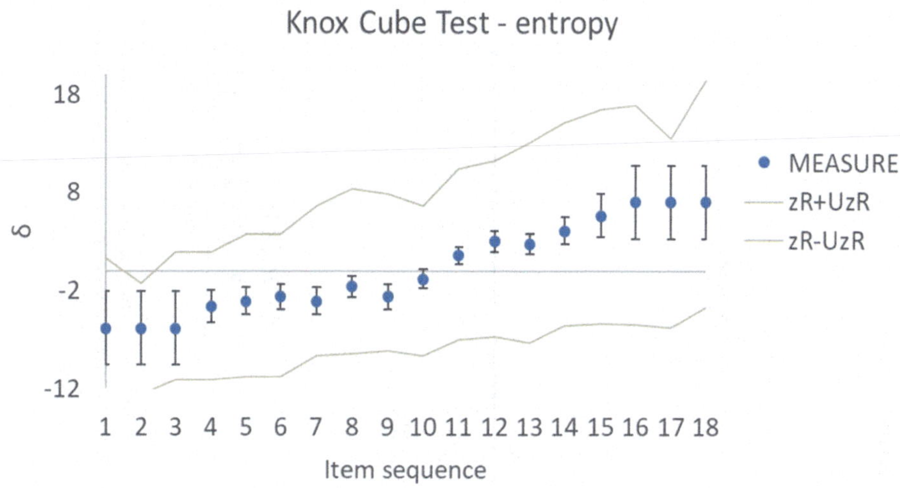

Fig. 5.6 Predicted values of task difficulty, δ, Eq. (5.10) for series of KCT sequences of increasing difficulty from a CSE, zR, Eq. (5.11) based on entropy Eq. (5.6) as well as reversals and average distance, compared with corresponding measurement values (blue dots with uncertainty intervals). The corridor of model uncertainties is shown as $\boldsymbol{zR} \pm \boldsymbol{UzR}$, $k = 2$ Eq. (5.12) (from Melin et al. 2019)

$$\begin{aligned}\delta_j &= zR_j \\ &= -9\ (5) + 2\ (1) \cdot \text{Entropy}_j + 0,8\ (1,4) \cdot \text{Reversals}_j \\ &\quad + 0,7\ (2,9) \cdot \text{AveDistance}_j \end{aligned} \tag{5.11}$$

5.3.1 *Measurement Uncertainty in Principal Component Regression*

Apart from the mean, $\hat{\theta}$, and standard uncertainty, $u(\theta)$, of the attribute value for each person (or item), it is of interest to express corresponding means and standard uncertainties in the regression coefficients of the construct specification equation relating the attribute value to a set of explanatory variables ('manifest predictors', $\boldsymbol{X}$), as well as statistics for significance testing of various differences among attribute values. This is summarised in Fig. 5.7 which shows an Ishikawa diagram for the general case.

The measurement uncertainties in the psychometric attribute values from Eq. (1.1) will propagate through the principal component regression described in

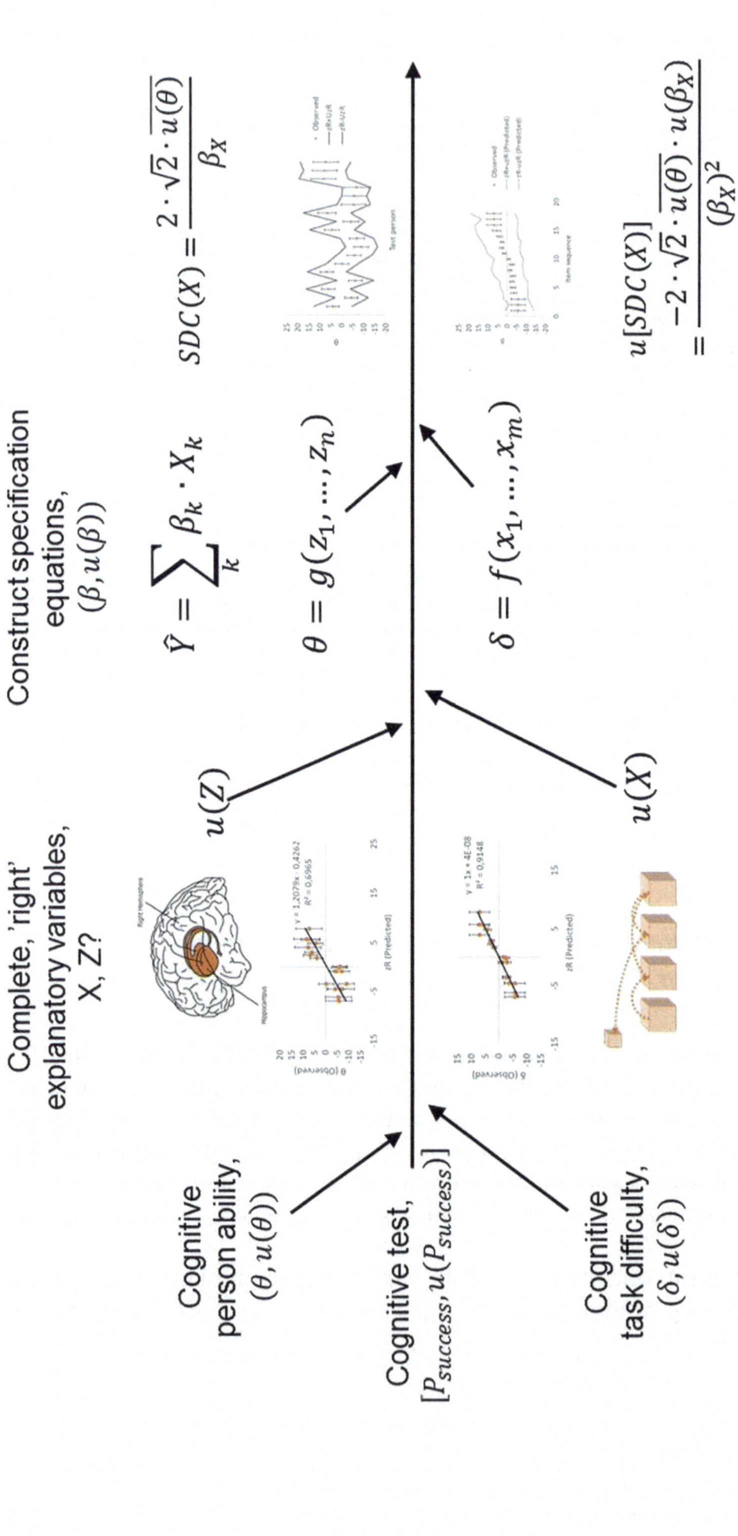

Fig. 5.7 Propagation of measurement uncertainties, portrayed as an Ishikawa diagram, from the initial psychometric cognitive test, through Rasch analysis; formulation of CSE for both cognitive ability (upper half) and task difficulty (lower half); and finally estimation of smallest detectable change (SDC) for each explanatory variable and Rasch attribute) (from Melin et al. 2019)

Sect. 5.2.5. An initial set of uncertainties in the estimates, $\hat{\boldsymbol{C}}$, of the coefficients from the present least-squared analyses Eq. (5.8) for the Knox cube test is $\hat{\boldsymbol{C}} = \begin{bmatrix} 2.0 & 0.3 \\ -0.5 & 1.4 \\ 0.7 & 3.2 \end{bmatrix}$, where the second column indicates the (expanded, $k = 2$) uncertainties in each coefficient (first column) of the multivariate model, as well as an 'intercept' value of—9.0(4.8) logits for the three principal components of variation for task difficulty, δ, and three explanatory variables: entropy Eq. (5.6); number of reversals and average 'distance' between the tapped blocks in each KCT sequence, respectively.

5.3.2 *Measurement Uncertainty in the Construct Specification Equation*

The above-mentioned uncertainties, $\boldsymbol{u}\left(\hat{\boldsymbol{C}}\right)$, in the least-squared coefficient estimates will propagate to produce uncertainties in the linear predictor (construct specification equation) $\hat{Y} = \sum_k \beta_k \cdot x_k$ Eq. (5.7), where the CSE coefficient expression $\boldsymbol{\beta} = \boldsymbol{P} \cdot \hat{\boldsymbol{C}}$ is used when transforming back from PCs $\boldsymbol{P}$ to the original explanatory variables $\boldsymbol{X}$, as follows:

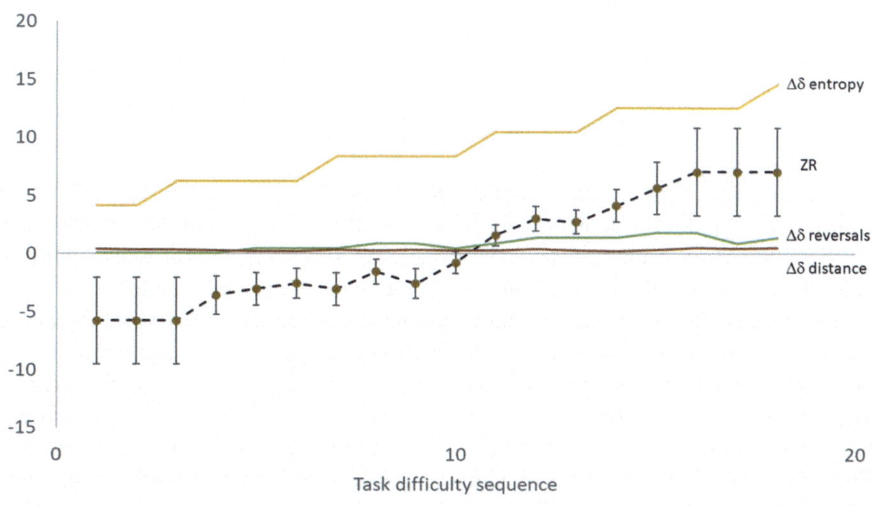

Fig. 5.8 Predicted contributions, $\Delta\delta$, to task difficulty Eq. (5.11) from the three explanatory variables $\boldsymbol{X} = \{\textbf{Entropy}, \textbf{Reversals}, \textbf{Average Distance}\}$ for the series of KCT sequences of increasing difficulty from a CSE, zR, based on entropy Eq. (5.6) (from Melin et al. 2019)

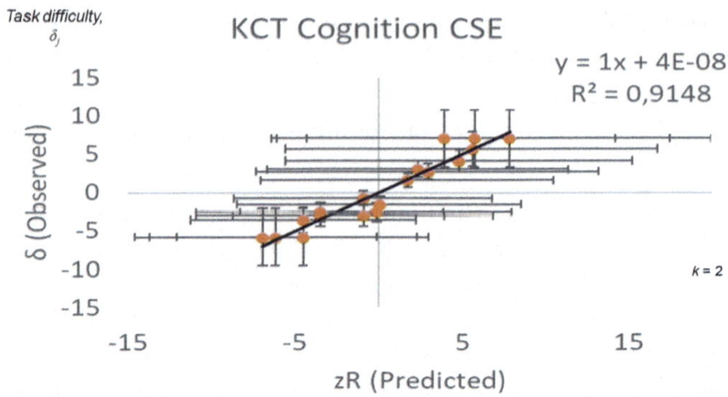

Fig. 5.9 Linear regression of the measured task difficulty, δ, against the CSE estimates zR based on the three explanatory variables $\boldsymbol{X} = \{\textbf{Entropy}, \textbf{Reversals}, \textbf{Average Distance}\}$ and Eqs. (5.6) and (5.11) for the series of KCT sequences (from Melin et al. 2019)

Firstly, $\boldsymbol{u}(\boldsymbol{\beta}) = \boldsymbol{P} \cdot \boldsymbol{u}(\hat{\boldsymbol{C}})$, where typical values for the present KCT case are $\boldsymbol{\beta} = \begin{bmatrix} 2.0 & 1.2 \\ 0.8 & 1.4 \\ 0.7 & 2.9 \end{bmatrix}$, where the second column indicates the (expanded, $k = 2$) uncertainties, $\boldsymbol{U}(\boldsymbol{\beta})$, in each coefficient (first column), as well as an 'intercept' value of −9.0(4.8) logits for the task difficulty, δ, in terms of the three explanatory variables: entropy ln *(G!)* Eq. (5.6); number of reversals and average 'distance' between the digits in each DST sequence, respectively.

Secondly, corresponding uncertainties in the linear predictor (construct specification equation) $\hat{Y} = \sum_k \beta_k \cdot x_k$ will be given by the combined (standard, $k = 1$) uncertainties:

$$u(\hat{Y}) = u(zR) = \sqrt{\sum_k u(\beta_k)^2 \cdot (x_k)^2} \tag{5.12}$$

The corridor of model uncertainties $zR \pm UzR$ ($k = 2$) shown in Fig. 5.6 for our entropy-based model is considerably wider than the measurement uncertainties $u(\delta)$, ($k = 2$) shown on each measure (see Fig. 5.2a). One explanation is that there are additional components of variation not yet included in the CSE model.

The results shown in Fig. 5.8 indicate that the explanatory variable '*Reversals*' has a small but slightly increasing contribution to the construct specification equation with increasing sequence difficulty. Figure 5.8 also indicates that the explanatory variable '*Average Distance*' contributes negligibly to the CSE compared with measurement uncertainties and can be safely eliminated without loss of information. The sole explanatory variable of significance appears to be '*Entropy*', calculated with Eq. (5.6). The overall ability of the construct specification equation (Eq. 5.11) to predict KCT task difficulty is indicated in Fig. 5.9 where observed and predicted values are plotted against each other.

In a comparable fashion to the Knox cube test (KCT), specification equations for task difficulty as a function of entropy can be developed for other sequence memory tests, such as the Corsi block test (CBT), the Digit Span Test (DST) of number sequences and the word lists of the Auditory Verbal Learning Test (AVLT). It is in fact found that the construct specification equations for several of these classic cognitive memory tests are in fact very similar, with substantially the same number of explanatory variables with similar coefficients Pendrill et al. (2019).

5.4 B: Instrument Construct Description and Specification

Continuing the passage of information Fig. 5.4, in this section the role of the instrument at the heart of the measurement system is now considered and how the concept of entropy can assist in describing, predicting and prioritising the instrument construct.

In considering the requirements for quality measurements in both the physical and social sciences Roach (2006), the validity of the instrument construct is almost as important as the initial entitic construct (Sect. 5.2). A second concept particularly relevant when considering the instrument is the responsiveness—i.e. the ability of an instrument to detect changes in the characteristic of interest.

The processes of prioritisation and structuring presented (Sect. 5.2.1) when formulating the entity construct can be followed analogously in the present case of formulating the instrument construct.

5.4.1 Instrument Attributes, Construct Specification and Entropy

We have already shown how changes in stimulus can be described and even explained in terms of the entropy associated with the entity, *H(P)* (Sect. 5.2). Turning now our attention to the instrument, a loss of information in a poorly performing measurement system can be described and even explained in terms of an increase in entropy, i.e. disorder, such as loss of a pattern as can be included in the entropy term $H(P, Q)$ in Eq. (3.3). Such considerations can inform when considering a valid description of the instrument.

The instrument is at the heart of the measurement system, thus playing a pivotal role so to say midway between the stimulus and the response.

In describing and explaining instrument entropy, the material presented in the earlier chapters is brought together:

- Most of the characteristics of measurement given in Table 2.4—such as resolution and modifying and interfering environmental effects—involve changes in either bias, b, or sensitivity, K, associated with the instrument, as described by Eqs. (2.5) and (2.6) for the processes of sensing; signal conditioning; signal processing; data presentation and decision-making.

- Known bias and known sensitivity of the instrument are of course corrected for. But if time and resources for measurement are limited (as they mostly are), there will be unknown bias and unknown changes in sensitivity. In Chap. 4, procedures for evaluating the measurement uncertainty arising from these unknowns were presented. For example, Fig. 4.4 can be used when evaluating uncertainty from the finite resolution of an instrument.
- In Sect. 4.4, the concepts of entropy, perception and decision-making were summarised.

An example of decision-making from the example of Man as a measurement instrument is elementary counting case, where the risk of decision errors when resolving two adjacent stimuli, s_a and s_b,—e.g. counting nine dots when there are ten (Pendrill and Fisher Jr 2015, Sect. 4.4.1)—is given by the overlap of the two (assumed) Normal distributions $N(s_a, (w \cdot s_a)^2)$ and $N(s_b, (w \cdot s_b)^2)$:

$$\alpha = \frac{1}{2} \cdot erfc\left(\frac{|s_a - s_b|}{w \cdot \sqrt{2\left(s_a^2 + s_b^2\right)}}\right) \tag{5.13}$$

where erfc is the complementary error function. The width of each distribution appearing in Eq. (5.13) is according to Weber's concept of the 'just noticeable difference' (JND), that is, the perceptual uncertainty, $w \cdot s$ of each observation, is the Weber constant w times the level of stimulus, s. Note in this context that the logarithmic form of the GLM (Sect. 3.5.1, Eq. (3.1)) is more general and is not related to the Weber–Fechner law of perception which applies in the special case of perception being proportional to the stimulus level (Pendrill and Fisher Jr 2015). Secondly, note that the JND is conceptually distinct from the 'minimum important change' (MIC) which relates to the *impact* of the distance instead (Sect. 6.3.2).

The conditional entropy expression:

$$H(Q|P) = H(P,Q) - H(P) \tag{3.3}$$

is equated in Eq. (4.4) with the subjective distance D_{KL}, between two distributions, P and Q, to be classified. There are many diverse examples of entropy-related changes in instrument performance.

One can be found in recent neurological research, with a study of the entropy of interconnectedness among different regions of the brain (Yao et al. 2013): the greater the order in brain processes, the smaller the entropy and the greater the (instrument) ability, for example, to memorise. In fact, these concepts can be applied to a myriad of 'item: probe' systems illustrated in Fig. 4.4, such as explaining the efficiency of a hospital in terms of the degree of order arising from synergy between different parts of the organisation (Chen et al. 2015) (Fig. 5.10).

An explanation of the ability, θ, of an instrument (e.g. person) analogous to our explanation of task difficulty is in terms of an entropy term $\theta = H(P,Q) = -\ln(G!)$, where G is the number of 'coherent connexions' between different parts of the instrument.

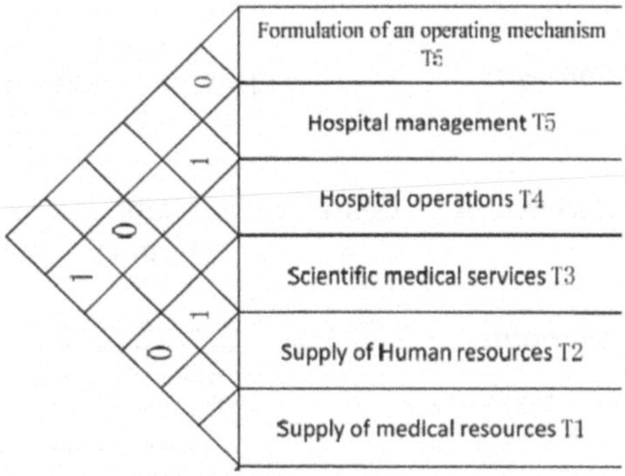

Fig. 5.10 Example of collaborative matrix indicators (Extract reproduced with permission. Chen et al. 2015)

Another term is the entropy $H(P, Q) \sim \ln(\rho) = \ln\left(\sqrt{3} \cdot 2 \cdot u\right)$ of a uniform distribution (Fig. 4.4) associated with the finite resolution, ρ, of an instrument, where u is the standard measurement uncertainty.

The unification of standard uncertainty, u, to decisions risks—as summarised in Eq. 5.2):

$$u_q \sim \mathrm{e}^{\Delta H(Q)} \tag{5.2}$$

can be applied to the present description of the instrument, where the case of uncertainty associated with the finite resolution of ρ can be continued. The sum of these instrument-related entropies increases the initial stimulus entropy $H(P) = \delta$, so that the resulting response, P_{success}, is similar to what is described by the 2PL IRT expression (Partchev 2004):

$$z_R = S = \theta - \delta = \ln\left[\frac{P_{\text{success}}}{1 - P_{\text{success}}}\right] - \ln(\rho) \quad (2.11 \text{ with } b = 0)$$

5.4.2 *Multi-Attribute Alternatives: House of Quality*

An example of structuring and prioritising (Sect. 5.2.1), from the extensive studies performed over the years by the OMERACT initiative in healthcare (Boers et al. 2018), shown in Fig. 5.11, illustrates the many faceted considerations, ranging from objective body structure (such as the skeleton) to more subjective, but no less important, aspects such as impact of the quality of daily living and family relations.

Similar diagrams and mapping of constructs can be done analogously in other fields. For instance, if one is concerned about the 'health' of an organisation rather than a person, then other indicators replace human measures with organisational factors, such as efficiency or effectiveness. Often these are related: understanding how the work environment affects employees and not just profits can improve not only the employees' health but also the health of the organisation (Heerkens et al. 2004).

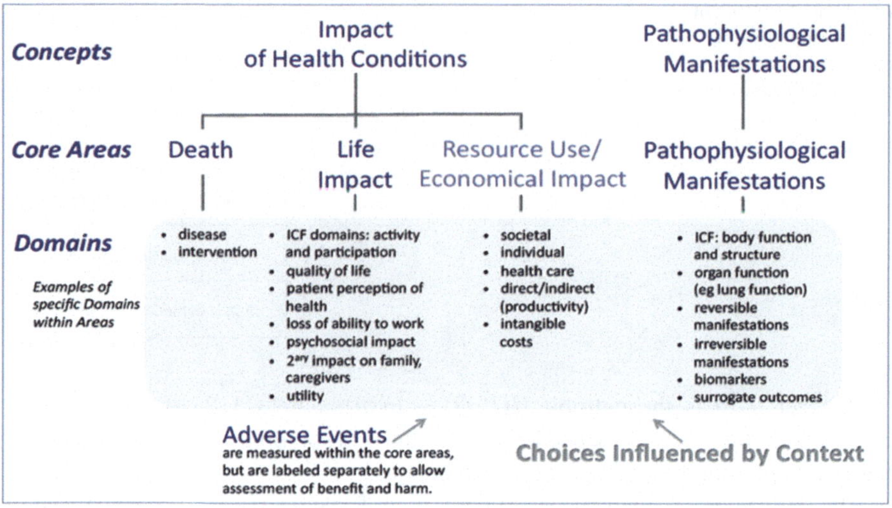

Fig. 5.11 Concepts, core areas and domains for outcome measurement in health intervention studies (Reproduced with permission of OMERACT, Boers et al. 2018)

Further examples of structural models can be found in the 'House of Quality' approach, popular in Quality Function Deployment (QFD, Akao and Mazur 2003; Chan and Wu 2002). Examples include models in which:

- system performance (e.g. efficiency of a hospital in terms of queue times, for instance) is related to process, cost and technical attributes (e.g. respectively, staff team building, personnel costs, and supply of medical resources) (Chen et al. 2015),
- the robustness of an organisation (e.g. to a terrorist attack or to human error) as determined by properties of components of a preventative or mitigating (security or quality assurance) system (Bashkansky and Dror 2015).

Relations between these responses and corresponding explanatory variables are visualised by the 'roof' of the House of Quality (or Security), which indicates the degree of correlation or importance between response and explanation.

Organisational Management Collaborative Entropy

An example where entropy can be used to express the degree of order in an organisation is in terms of the number X_{ijq} of fully collaborative work stages j in a division i where the total number X_{ij} also includes a number X_{ijf} of non-collaborative stages, such that the collaborative entropy for organisational management is given by (Chen et al. 2015):

$$H_{ijq} = -\frac{X_{ijq}}{X_{ij}} \cdot \log\left(\frac{X_{ijq}}{X_{ij}}\right); X_{ij} = X_{ijq} + X_{ijf} \tag{5.14}$$

The smaller the entropy, the more coordinated (and more efficient) the organisation is.

As illustrated in the above figure (Fig. 5.10), X_q is the number of scores '1' where collaboration occurs between two work stages (j, grouped by processes, costs, technical), while X_f is the number of scores '0' where collaboration does not occur. At one level of an organisation, the level of collaboration can be expressed as:

$$\mathrm{CD} = 1 - \frac{H_{ijq}}{\max H_{ijq}} \tag{5.15}$$

in relation to the largest collaborative entropy, $\max H_{ijq}$, at that level. The corresponding collaborative efficiency is

$$\mathrm{EF} = 1 - \frac{H_{ijq}}{H_{ijq} + H_{ijf}} \tag{5.16}$$

The integrated collaborative entropy for the organisational management elements at all levels is

$$B = \sum w_i \cdot H_{ij} \tag{5.17}$$

as the sum over the elements, weighted by the importance w_i of each element i.

The advantages of being able to sum arithmetically the components of entropy, as noted in Sect. 5.1.2, are clear from the formulations (5.14)–(5.17).

Synergy

A second approach to expressing collaborative efficiency is to score positive (independent, negative) synergy between two work stages (j, j') (called 'HOWs', i.e. prevention/mitigation tools, by Bashkansky and Dror (2015)) as $\Delta_{j,j'} = +1$ ($0, -1$), where the joint effect is greater (equal, less) than the sum of the parts acting alone. A synergy factor for stage j is calculated as a sum over all other stages:

$$s_{i,j} = 1 + \frac{\sum_{j \neq j'}^{J} \Delta_{i,j,j'}}{J - 1} \tag{5.18}$$

Our approach would be to regard the ability of the organisation, in particular the benefits of synergy, in terms of the entropy of the 'probe' in tackling a task of a certain level of difficulty.

Finally, the above expressions can not only be used to describe the status quo, for instance, in the current efficiency of an organisation, but also can be deployed when determining measures of the impact of various interventions aimed at improving that efficiency. An example is the deployment of state-of-the-art localisation systems in which instrumentation, personnel or patients are tagged in, say, the emergency department of a hospital (Fisher and Monahan 2012; Christe et al. 2010). The effects of such an intervention are measured not only in terms of the localisation accuracy or battery life, but also in terms of reducing unnecessary overinvestment in equipment and, perhaps most importantly, freeing resources so that personnel can dedicate more of their time to patient care instead of searching for lost equipment.

5.4.3 *Formulation of Instrument Construct Specification Equations*

A person (probe or instrument) construct specification equation can be formulated in a very similar way to the item construct specification equation, Eq. (5.8), as:

$$\theta_i^* = -\left(\sum_{k'} \beta_{ik'} \cdot x_{k'} + \varepsilon_\theta\right)$$

where $x_{k'}$ denotes the explanatory variable of cognitive ability k', and $\beta_{i,\ k'}$ is the weighting of each cognitive ability component (e.g. volume of amygdala, concentration of neurofilament light in CSF, level of education, etc.) for each person (or instrument) i. A fit of the data to the model can be examined by comparing Rasch model estimates of person abilities, θ_i, with those estimated by calculating difficulties from the component ability estimates, θ_i^*, using a maximum likelihood statistic, such as in Eq. (5.12).

5.5 C: Response, Error and Entropy—Categorical Observations

Finalising the passage of information, in this section the concept of entropy can assist in describing, predicting and prioritising how the rater judges the response and performs final restitution of the measurand from the response (Fig. 5.2).

In considering the requirements of validity and reliability for measurements in both the physical and social sciences (Roach 2006), much emphasis when considering rating, response and restitution will be particularly on

- *Reliability*: Outcome measure produces the same number each time instrument is used.

- *Self-reporting*: Test–Retest reliability, e.g. limited by wording and interpretation.
- *Internal consistency*: Do all items in outcome measure address same underlying concept?
- *Rater performance*: Intra-rater consistency with repeats + inter-rater consistency among different raters.
- *Reliability coefficient*: True score variance/(true score variance + error variance).

As described in Chap. 2, distortion in the more general case including ordinal and nominal data arising somewhere in the measurement system can be expressed as:

- Accuracy (decision-making) = response categorisation – input (true) categorisation (2.8)

Categorical scales (introduced in Sect. 1.2.3), it is recalled, are often invoked in situations where, for often practical reasons and limitations (much like causes of measurement uncertainty), one chooses to merely classify responses into a finite number of categories rather than attempt a continuous quantification. Continuing our presentation in Sect. 5.1.1, here we consider interhistogram distances between probability mass functions (PMF) in the response of a measurement system. Thereafter we will make the transition from discrete, categorical scales to the more familiar continuous scales of quantitative data.

5.5.1 Interhistogram Distances on Categorical Scales: Systematic Errors

There is an extensive literature about different measures of the distance between histograms with a myriad of applications including image recognition. Pele and Werman (2010) studied a number of distance metrics applied to a simple set of histograms, shown in Fig. 5.12. In image recognition, the height of a histogram column for a certain category could be a measure of the occupancy (that is, how many images) containing a particular image feature (be it colour, sharpness, size) or whatever quality characteristic of the entity.

In comparing the performance of these various histogram distance measures, aspects such as computational expense needed to calculate them are a consideration as well as the fact that each measure gives a different estimate of the dissimilarities of the histograms (Pele and Werman 2010; Rubner et al. 2000). As reviewed by those authors, there is as yet no unique measure of interhistogram distance which works satisfactorily in the general case, as is evident to anyone who has attempted to match images over the Internet, where tools such as Google Image® use the algorithms referred to here: each algorithm mentioned by Pele and Werman (2010 and references therein) produces different results for the same data.

A typical case is illustrated in Fig. 5.12 where one is interested in measures of the distances separating pairs of histograms, in particular establishing a measure of which of the three responses (a), (b) or (c) $P(y|z)$ lies closest (i.e. is the most faithful

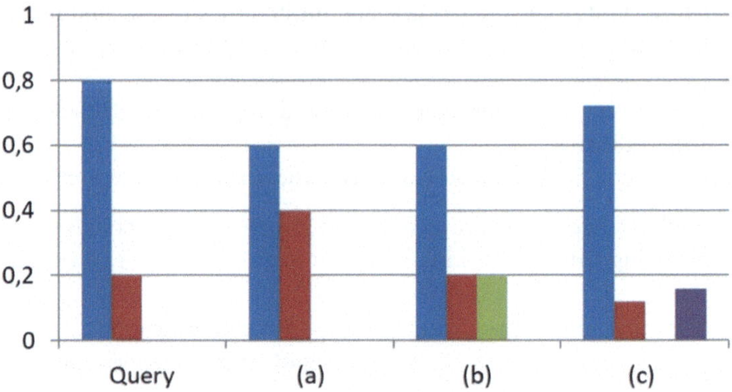

Fig. 5.12 PMF histograms of the original 'Query' and three experimental classifications (a), (b) and (c), each distributed over four categories (adapted from Pele and Werman (2010))

response, y) to the original stimulus $P(z)$, labelled 'Query'. In the example shown, the occupancy, $\boldsymbol{z}$ and $\boldsymbol{y}$, of up to $K = 4$ available categories, $\boldsymbol{c} = \{\varepsilon_1, \varepsilon_2, \varepsilon_3, \varepsilon_4\}$ varies in each of samples Z and Y being pairwise compared, where, for example, the response PMF $Y = P(y|z)$ is compared with the 'Query' PMF $Z = P(z)$.

Pele and Werman (2010) investigated several different histogram distance metrics, including:

- $\boldsymbol{X}^2$ on a *syntax* scale: $\chi^2(Z, Y) = \frac{1}{2} \cdot \sum_c \frac{(z_c - y_c)^2}{z_c + y_c}$,
- the Quadratic Form $QF^A(Z, Y) = \sqrt{(Z - Y)^T \cdot A \cdot (Z - Y)}$. When the bin-similarity matrix $\boldsymbol{A}$ is the inverse of the covariance matrix, the Quadratic-Form distance is the Mahalanobis distance (Pele and Werman 2010 and references therein),
- an entropy-based measure of interhistogram distances on a *semantic* scale which is a variant of the Kullback–Leibler (1951) distance (Eq. 4.5):

$$d_{\mathrm{KL}}(Z, Y) = \sum_c z_c \cdot \ln_b \left(\frac{z_c}{y_c} \right)$$

where for the examples shown in Fig. 5.11, the log base is $b = 4$, equal to the number, C, of categories.

As is well-known, the Kullback–Leibler (1951) distance is not considered a complete metric since in general it is not symmetric, i.e. $d_{\mathrm{KL}}(Z, Y) \neq d_{\mathrm{KL}}(Y, Z)$ (Sect. 5.5.3). Since the KL distance is not strictly a metric, alternative measures are sought where the variant of d_{KL} after Jeffrey is one (Pele and Werman 2010 and the references therein):

$$d_J(Z, Y) = \sum_c \left[z_c \cdot \ln\left(\frac{z_c}{m_c}\right) + y_c \cdot \ln\left(\frac{y_c}{m_c}\right) \right]$$

where $m_c = \frac{z_c + y_c}{2}$.

In its infinitesimal form the Kullback–Leibler (1951) distance is a metric tensor: the Fisher information metric, $g_{j,\,k}(\theta_0)$ which appears to second order in a Taylor expansion of the Kullback–Leibler distance, on small displacements $\Delta\theta^j = (\theta - \theta_0)^j$:

$$d_{\mathrm{KL}}(P(\theta)|P(\theta_0)) = \Delta\theta^j \cdot \Delta\theta^k \cdot g_{j,k}(\theta_0) + \ldots$$

The Fisher information metric, as a Hessian matrix of the divergence,

$$g_{j,k}(\theta_0) = \frac{\partial^2}{\partial\theta^j \cdot \partial\theta^k}_{\theta=\theta_0} d_{\mathrm{KL}}(P(\theta)|P(\theta_0))$$

then enters, for example, in the Wald test statistic $z = \frac{\hat{\theta} - \theta_0}{\mathrm{SE}}$, where $\mathrm{SE} = \frac{1}{\sqrt{\frac{\partial^2}{\partial\theta^j \cdot \partial\theta^k}_{\theta=\theta_0} d_{\mathrm{KL}}(P(\theta)|P(\theta_0))}}$. In terms of entropy $(z, \theta) = -\ln\,[p(z, \theta)]$, $g_{j,k}(\theta_0) = \int \frac{\partial^2}{\partial\theta^j \cdot \partial\theta^k}_{\theta=\theta_0} H(z, \theta) \cdot p(z, \theta) \cdot dz = \mathbb{E}\left[\frac{\partial^2}{\partial\theta^j \cdot \partial\theta^k}_{\theta=\theta_0} H(z, \theta)\right]$. In Eq. (4.4), we set an equivalence between the subjective distance $D_{\mathrm{KL}}(a, b)$ and the conditional entropy $H(Q|\,P)$.

5.5.2 *Response and Entropy*

Measurement uncertainty is some estimate of the 'unknown' errors which one has not managed to evaluate and correct for because of limited time and resources and captures how well the observer's expectations match the output of the measurement system.

Intrahistogram Distances: Measurement Uncertainty

Measures of the dispersion within a particular PMF are needed when attempting to describe measurement uncertainty, as a complement to the expressions for the distance between different PMFs (dealt with above). Bashkansky and co-workers have over the years proposed various expressions of variation applicable to nominal and ordinal properties.

Bashkansky and Gadrich (2010), considering classification errors in terms of the probabilities of incorrect decisions $P_{k,k'}$ (off-diagonal elements of the confusion matrix $\boldsymbol{P}$ (Sect. 2.4.4) which are measures of the occupancy $\boldsymbol{c}$ of the response PMF), describe the uncertainty, U, in a classification in terms of the likelihood that a classification k' is made, whereas the true classification is k, which from Bayes' theorem gives

$$U_{k,k'} = \frac{p_k \cdot P_{k,k'}}{\sum_{k=1}^{K} p_k \cdot P_{k,k'}}$$

Gadrich et al. (2014) proposed a generalisation of the so-called Gini formula (used frequently in econometrics and the social sciences), where a total (population) variation is defined by

$$V_T = \sum_{k=1}^{K} \sum_{k'=1}^{K} L(c_k, c_{k'}) \cdot p_k \cdot p_{k'}$$

where the 'loss-of-similarity' function, L, can take various expressions depending on the underlying scale. For a nominal scale $L(c_k, c_{k'}) = \begin{cases} 0 \text{ when } k = k' \\ 1 \text{ when } k \neq k' \end{cases}$, and for an ordinal scale, $L(c_k, c_{k'}) = |k - k'|$, where this distance metric is based on counting the number of steps needed to connect the categories c_k and $c_{k'}$ (see also Mencattini and Mari (2015)).

As remarked in Chap. 2, in example 1 of VIN §3.9 *Examination uncertainty* given by Nordin et al. (2018): 'The reference nominal property value is "B". The nominal property value set . . . of all possible nominal property values is {A, B}. For one of 10 examinations . . . the examined value differs from "B". The examination uncertainty is therefore 0.1 (10%)'.

Again misclassification probabilities, α and β, such as the 10% mentioned here are considered as accuracy measures, but such performance metrics often belong to the 'counted fraction' kind of data and in general are of ordinal, rather than fully quantitative nature, not directly amenable to regular statistics (Sect. 3.5.1). There is also the same task of separating the instrument factor from the sought-after object factor which needs to be done even for qualitative, categorical responses of the measurement systems. These various expressions of measurement uncertainty suffer generally from the same limitations on ordinal or nominal scales as those encountered for interhistogram distance measures (above) and similarly can be better treated in terms of entropy.

In many cases the 'nominal' aspect encountered in measurement is not an intrinsic property of the measured (or classified/examined) entity, but is rather a characteristic of the response of classification system employed. For often practical reasons and limitations (much like causes of measurement uncertainty) one chooses to merely classify responses into a finite number of categories rather than attempt a continuous quantification (Fig. 1.2). The morphology of blood cells can be expressed in continuous, quantitative terms, but is often measured (classified) in a finite set of discrete and nominal categories of response. The 'examinand' can however be recovered as a continuous property in the process of restitution from the classification response. This is the subject of the next section.

5.5.3 Deriving Response

It is straightforward and instructive to recall how the logistic transformation of counted fractions (Sect. 3.5.1) can be derived from first principles. In fact, the probabilities, q_c, of classifying a response in a category c, used in our process of restitution (Fig. 3.3), can be modelled by deriving decision probabilities with the method of Lagrange multipliers subject to the constraint of maximising entropy, as follows (Pendrill 2017).

Consider a random variable $\boldsymbol{y} = (y_1, y_2, \ldots, y_C | y_c \in \mathbb{R}^{+})$ associated with the response of the measurement system (e.g. with Man as the measurement instrument, although the discussion is quite generic) to a stimulus, z, from the quality characteristic of the entity, such as the level of difficulty of a task. Assume that the stimulus values attributed to the entity are themselves, prior to measurement, distributed over a range of categories k, with a priori probability p_k, as described in Sect. 5.5.2.1.

The probability distribution of the corresponding responses, $\boldsymbol{y}$, of a measurement system to the stimulus values, z, which needs to be derived is $\boldsymbol{q} = [q_1, q, \ldots, q_C]$, that is, the PMF $P(y_k| z_M)$,given we know the mean (expected) value $\mathbb{E}(\boldsymbol{y})$ given by Eq. (5.5): $\mathbb{E}(y|z_M) = \sum_{k=1}^{K} \frac{P(y_k|z_M) \cdot y_k}{K}$.

The concept of entropy is key in deriving these responses on all measurement scales, including the ordinal and nominal. In an analogous fashion to explaining task difficulty (e.g. of a memory test item) in terms of entropy (Sect. 5.5.2.1), the distribution of responses, which is determined by how well an instrument (person) makes the correct decisions of choice and identification ((e.g. Eq. (2.6) in Sect. 2.4.3), will depend in some way on the degree of order in the instrument processes which is expressed in terms of entropy.

According to the principle of maximum (Shannon) entropy (Sect. 3.2.2), the change in entropy on transmission of measurement information from stimulus through response of the measurement system cannot decrease. Applying this principle here means maximising the entropy function in response, that is, after transmission:

$$H[q_1, q, \ldots, q_C] = -\sum_{c=1}^{C} q_c \cdot \ln(q_c)$$

subject to the constraints: $q_c \geq 0; \sum_{c=1}^{C} q_c = 1$ and the expected response $\mathbb{E}(\boldsymbol{y}) = \sum_{c=1}^{C} q_c \cdot y_i$

The Lagrange function is

$$L = -\sum_{c=1}^{C} q_c \cdot \ln(q_c) - \alpha \cdot \left(\sum_{c=1}^{C} q_c - 1\right) - \rho \cdot \left[\sum_{c=1}^{C} q_c \cdot z_i - \mathbb{E}(\boldsymbol{y})\right]$$

with corresponding respective partial derivatives set equal to zero:

$$\frac{\partial L}{\partial q_c} = -\ln(q_c) - 1 - \alpha - \rho \cdot y_c = 0$$

$$\frac{\partial \mathrm{L}}{\partial \alpha} = 1 - \sum_{c=1}^{C} \mathrm{q_c} = 0; \quad \frac{\partial L}{\partial \rho} = \mathbb{E}(\boldsymbol{y}) - \sum_{c=1}^{C} q_c \cdot y_i = 0$$

where the latter two expressions describe the variations close to maximum entropy with respect to the Lagrange multipliers α and ρ corresponding to the two constraints.

Each categorical probability, q_c, of response is then given from the first expression as:

$$q_c = e^{-1-\alpha-\rho \cdot y_c} = e^{-(1+\alpha)} \cdot e^{-\rho \cdot y_c}$$

which, normalised to the sum of the probabilities, will be

$$q_c = \frac{e^{-\rho \cdot y_c}}{\sum_{c'=1}^{C} e^{-\rho \cdot y_{c'}}} \tag{5.19}$$

Equation (5.19) is indeed the transformation applied to the counted fraction distribution (Sect. 3.5.1) which is not surprising considering the first constraint and the associated second partial derivative of the Lagrange function with respect to α.

In the simplest dichotomous case, the response is the (equally weighted) random variable $\boldsymbol{y} = \{-0.5, 0.5 | y_k \in \mathbb{R}^+\}$ and its expectation is

$$\begin{aligned} \mathbb{E}(\boldsymbol{y}|z_M) &= \sum\nolimits_{k=1}^{K=2} \frac{P(y_k|z_M) \cdot y_k}{K} = -0.5 \cdot P_{\text{success}} + 0.5 \cdot (1 - P_{\text{success}}) \\ &= 0.5 - P_{\text{success}} \end{aligned} \tag{5.20}$$

In that binary case, Eq. (5.19) becomes the logistic Rasch expression Eq. (1.1) when the link function $\rho \cdot y_c = \theta - \delta$. From the corresponding 2PL model: $\rho \cdot y_c = \rho \cdot (\theta - \delta)$, where ρ is the discrimination.

Since this Lagrange multiplier derivation depends on maximising entropy, possible relations between entropy, the link function $z = \rho \cdot y_c$ in Eq. (5.19) and the response distribution are expected. Indeed, on inspection of the second constraint and the associated third partial derivative of the Lagrange function with respect to ρ, it becomes clear that close to maximum entropy:

$$\partial L = \partial H = \mathbb{E}(y) \cdot \partial \rho = \Delta P_{\text{success}} \cdot \partial \rho \tag{5.21}$$

where $\Delta P_{\text{success}}$ is a small change in the (average binary) response of the measurement system given by Eq. (5.19).

As already proposed in Chap. 2, in the simplest binary case of categorical responses of, for instance, the decision-making elements Eq. (2.8), the response can be modelled as $P_{\text{success}} = K \cdot z$ by analogy to the usual measurement system response.

Since the system response R is the product of instrument sensitivity, K, and stimulus, z, Eq. (5.21) appears to indicate that a change in entropy could be explained either in terms of a change in K or stimulus $\theta - \delta$. (These binary response expressions can be readily generalised to the polytomous case when needed.)

Further discussion of the relations between instrument sensitivity and ability can be found in Sect. 6.5.3.

5.6 Modelling Measurement System Response

5.6.1 Ordinary Factor Analysis

In qualitative measurement, a separation of the two attributes (probe/item) in the response of a measurement system, necessary among others for the establishment of metrologically traceable measurement, could in principle be attempted with an ordinary factor analysis in traditional statistics. As described by Wright (1994), a model of each measurement result, $z_{i,j} = \frac{x_{i,j} - x_{\cdot,j}}{s_{\cdot,j}}$, with score x, standard deviation s and mean x, for test person (TP) i with attribute (e.g. ability) y_i and item j with attribute (e.g. challenge) v_j could be

$$z_{i,j} = y_i \cdot v_j + \varepsilon_{i,j}$$

with measurement error ε, estimated by minimising:

$$\sum_{i=1}^{N_{\text{TP}}} \sum_{j=1}^{L} \left(z_{i,j} - y_i \cdot v_j\right)^2 \tag{5.22}$$

where L is the 'length' of the sample, i.e. the number of items. A recent example of such a traditional approach to separating person and item attributes is reported from

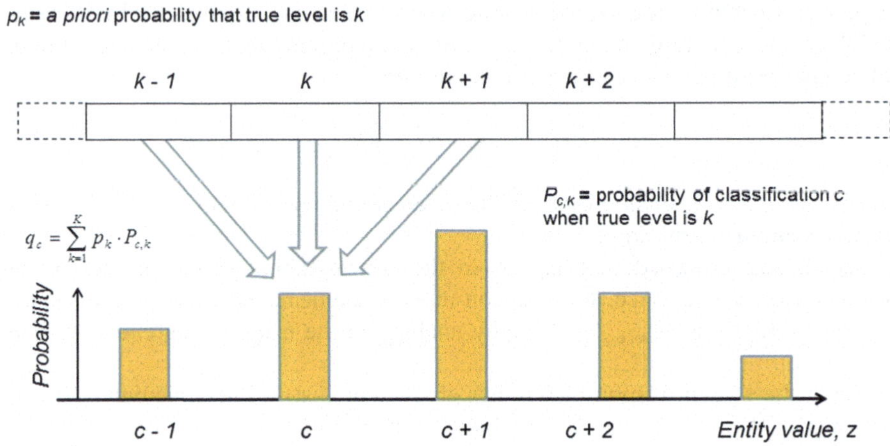

Fig. 5.13 Polytomous scale (Likert scale from Linacre 2002)

sensory science by Verhoef et al. (2015), who in turn refer to Cronbach et al.'s (1963) 'generalisability' theory.

Such approaches would however not necessarily work for ordinal data since the distance between each data point and the model estimate in expressions such as Eq. (5.22) is in most cases not referred to a fully quantitative scale (Svensson 2001, Chap. 3).

5.6.2 *Psychometric Factor Analysis*

In Chap. 4, we recalled how a logistic regression of the Rasch formula Eq. (1.1) to the experimental response data provides a widely accepted solution for handling ordinal data. But, as noted in the footnote in Sect. 4.4, the psychometric Rasch version of Eq. (1.1):

$$\sum_{i=1}^{N_{\mathrm{TP}}}\sum_{j=1}^{L}\left(y_{i,j} - P_{\mathrm{success},\,i,j}\right)^2 \tag{5.23}$$

also tacitly assumes that there is quantitative meaning which might be lacking in the distance $y_{i,j} - P_{\mathrm{success},\,i,\,j}$ between each experimental response, $y_{i,\,j}$ of the measurement system and the fitted value $P_{\mathrm{success},\,i,\,j}$ of the probability of successful classification (Fig. 5.13).

The residual $y_{i,j} = x_{i,j} - \mathbb{E}_{i,j}$ for the ith person (instrument) and j^{th} item (object or entity characteristic) is the difference between the score $x_{i,j}$ and the expected mean $\mathbb{E}_{i,j} = \sum_{k=0}^{m_i} k \cdot q_{i,j,k}$ for the polytomous case of the Rasch model where

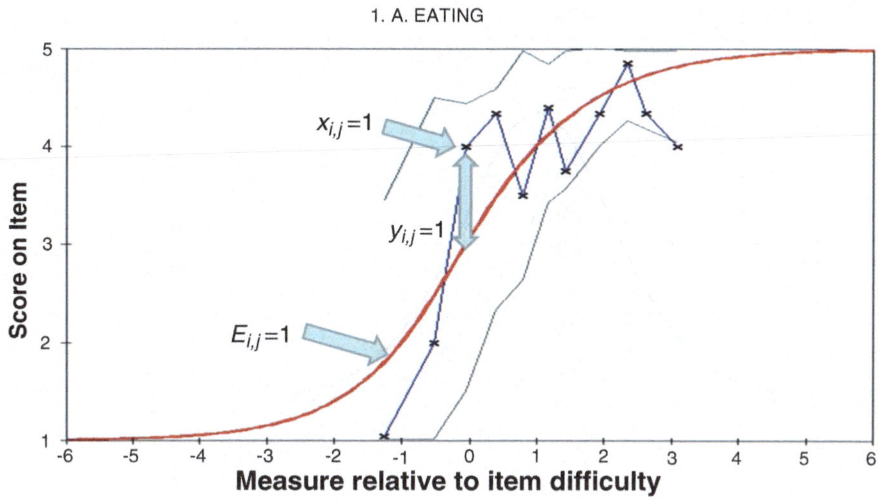

Fig. 5.14 Residuals of logistic regression as differences $y_{i,j} = x_{i,j} - \mathbb{E}_{i,j}$ between each observed score $x_{i,j}$ and the expected score $\mathbb{E}_{i,j}$ for one item (task of 'eating') and across the cohort of test persons, i, in the example data set 'EXAM12' provided with the WINSTEPS® program

the probability, q, of obtaining a response in classification category k is $q_{i,j,k} = \frac{e^{\sum_{c=0}^{k}\left(\theta_i - \delta_{j,c}\right)}}{\sum_{k=0}^{C_j} e^{\sum_{c=0}^{k}\left(\theta_i - \delta_{j,c}\right)}}$. These various statistics are indicated in Fig. 5.14 for a typical case (the data set 'EXAM12' provided with the WINSTEPS® program).

A common overall measure of the goodness-of-fit for the jth item is the chi-squared metric in the response of the N_{TP} test persons (or instruments):

$$\chi_j^2 = \sum_{i=1}^{N_{TP}} y_i^2 \tag{5.24}$$

Such quantitative meaning in these residual measures is in itself of course not always assured even when using the Rasch approach, for the very reason that one is treating ordinal data in the response of the measurement system, and the residuals are differences in success probabilities. We will review the various diagnostic tests routinely performed when verifying the reliability and validity of the Rasch approach to handling ordinal data later in this chapter. But in our opinion, the most correct approach to report and present of measurement results is best done, not with the response, but with the restituted estimates of the original attribute data associated with the measured entity, where the measurand which constitutes the stimulus applied at the input to the measurement system is treated on a quantitative scale (Fig. 5.4).

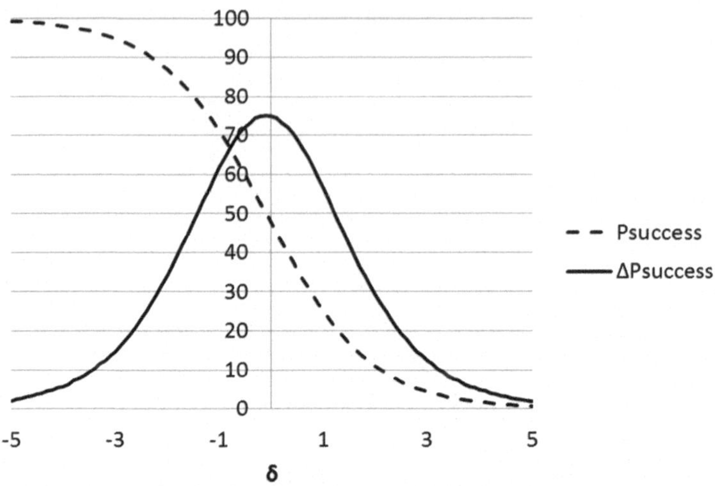

Fig. 5.15 Decision-making performance for a (binary) measurement system: Item response function (dashed line) and sensitivity (full line) calculated with Eqs. (1.1) and (5.25), respectively

5.6.3 Sensitivity of System for Ordinal Data

When choosing among the many tools traditionally used to check the Rasch model, it will be useful to invoke a particular model based on measurement system analysis (MSA, Fig. 5.4, Pendrill and Petersson (2016)), specifically where the measurement system has an output response which is a performance metric (i.e. how 'well' a task is performed). A key parameter is the sensitivity (K) of the system which may be readily derived as the change in response (R) to a change in stimulus (S) by simply differentiating the response P_{success} entering into the Rasch formula Eq. (1.1). For a given response P_{success} there might in general be several different values of the stimulus z or, alternatively, a certain value of stimulus z might have different scores for different sets of items or different groups of persons (Andrich 2017).

A salient feature of a measurement system described with the Rasch model is the strong non-linearity in the sensitivity of an instrument characterised in terms of decision-making performance, as illustrated in Fig. 5.15 (Pendrill and Petersson 2016). This is in strong contrast to the more or less constant, often close-to-one sensitivity of most engineered instruments in the 'hard' sciences (Bentley 2005, Chap. 2).

Our readily made calculations of sensitivity using the MSA model involve differentiating the system response with respect to changes in stimulus: $K = \frac{\partial P_{\text{success}}}{\partial \delta}$. Starting with the (dichotomous) Rasch formula, $P_{\text{success}} = \frac{e^{\theta-\delta}}{1+e^{\theta-\delta}}$ (Eq. (1.1); solid line in Fig. 5.15), the sensitivity, K, can be derived as:

$$K = \frac{\partial P_{\text{success}}}{\partial \delta} = \frac{e^{2 \cdot (\theta - \delta)}}{\left(1 + e^{(\theta - \delta)}\right)^2} - \frac{e^{(\theta - \delta)}}{1 + e^{(\theta - \delta)}} \tag{5.25}$$

and is plotted as the dashed line in Fig. 5.15.

5.7 Metrological Comparability and Uncertainty of Ordinal Data: Scale and Sensitivity Distortions

Both key aspects of metrology—comparability (through traceability) and uncertainty—in categorical measurements can be accommodated in the MSA framework by deducing the effects, respectively, of systematic distortions across the scale investigated and random noise on the response of our measurement system approach with performance metrics. As said, the appropriate scale for treating these metrological aspects is that of the measurand, deduced from a restitution process largely based on the Rasch transformation as a special case of the Item Response Theory.

The wider validity of the Rasch model in psychometrics continues to be a subject of debate—recommended readings include both a paper by Humphry (2011) and a series of comments in the same issue about his paper which contain many contemporary concerns about the discipline. The idea of unit definitions based on Rasch models, and so also the potential of Rasch measurement to support metrological unit traceability, is controversial. A word of caution (Fisher 2015) is that a good fit to a Rasch model does not in itself automatically confer properties of invariance, parameter separation, unidimensionality, etc. on scores or measures. 'In fact, the dimensional nature of the FIM is very well understood and described in many published papers. Perhaps, however, there is an opportunity here for some energetic analyst to investigate the FIM using different fit statistics. (Linacre 1996) *The Rasch model cannot be "disproved"!*'

Scepticism about the Rasch model often quotes the following (Dorans et al. 2010): 'The search for a single best model that could be employed universally would be unwise data analysis. As Tukey (1986) indicated in his discussion of Rasch's (1961) quest for the best fitting model, "...We must be prepared to use many models, and find their use helpful for many specific purposes, when we already know they are wrong—and in what ways. ...In data analysis...we must be quite explicit about the deficiencies of the models with which we work. If we take them at face value, we can—all too frequently—be led to unreasonable and unhelpful actions. If we try to make them 'fit the facts,' we can ensure sufficient mathematical complexity to keep us from any useful guidance" (p. 504)'.

There are a number of assumptions underlying the basic Rasch model which, of course, have to be tested for validity in each particular application. Assumptions include (Christensen and Olsberg 2013) that items should:

1. measure only one latent variable: θ is a scalar, unidimensionality

Items should deal with only one subject (e.g. not be a mixture of math and language items).

2. increase with the underlying latent variable: $\theta \to \mathbb{E}\left(P_{\text{success},j}|\theta\right)$ increases for all items, j, i.e. monotonicity

The probability of a correct answer should increase with ability.

3. be sufficiently different to avoid redundancy: local independence $p\left(\overline{X} = \overline{P_{\text{success}}}|\theta\right) = \prod_{j=1}^{N_{\text{items}}} p\left(X_j = P_{\text{success},j}|\theta\right)$ for all θ

Items should not ask the same thing twice. Christensen and Olsberg (2013) explain: 'since we would expect two similar items to be highly correlated, and even to have a higher correlation than the underlying latent variable accounts for, it is usual to impose the requirement of local independence'.

4. function in the same way in any sub population: $p(X_j = P_{\text{success},\ j}| Y,\ \theta) = p(X_j = P_{\text{success},\ j}| \theta)$ for all items, j, and all variables $\boldsymbol{Y}$. Absence of DIF (differential item functioning)

Again Christensen and Olsberg (2013): 'The difficulty of an item should depend only on the ability of the student, e.g. an item should not have features that make it easier for boys than for girls at the same level of ability'.

Differential item functioning (DIF) analysis examines item fit with respect to person-groups, particularly that a measure should work in the same way across groups. Without this, there would be no assurance that comparisons could be made between different groups. DIF is commonly examined with analysis of variance of fit residuals, again by inspection of the ogive curves.

Such approaches do not however automatically take account of the rapidly changing sensitivity of the measurement system response, as described in Sect. 5.6.3. Starting from Expression (5.24), a number of measures of goodness-of-fit have been formulated over the years to check how well regression of the Rasch model to a particular set of experimental data is achieved in each case.

As mentioned earlier in this book (Sect. 2.4), as with traditional engineering measurement systems, in the social sciences one will need in general to account for 'incorrect' behaviour of a person acting as a measurement instrument by including several of the measurement characteristics listed in Table 2.4, such as guessing, bias and variations in sensitivity (aka 'discrimination') (as dealt with in extensions to the basic Rasch model, such as the so-called 3PL logistic model (Partchev 2004).

There is an extensive literature about the design and analysis of questionnaires and surveys, including definitions and strategies to compensate for a number of common response strategies regularly encountered, perhaps more commonly in socioemotional investigations than in, for instance, cognitive studies (Soto et al. 2008; van der Bles et al. 2019). Among such effects (Fig. 5.16) are:

- Acquiescence: 'Agreement regardless of item content'.
- Disacquiesence: 'Disagreement regardless of item content'.

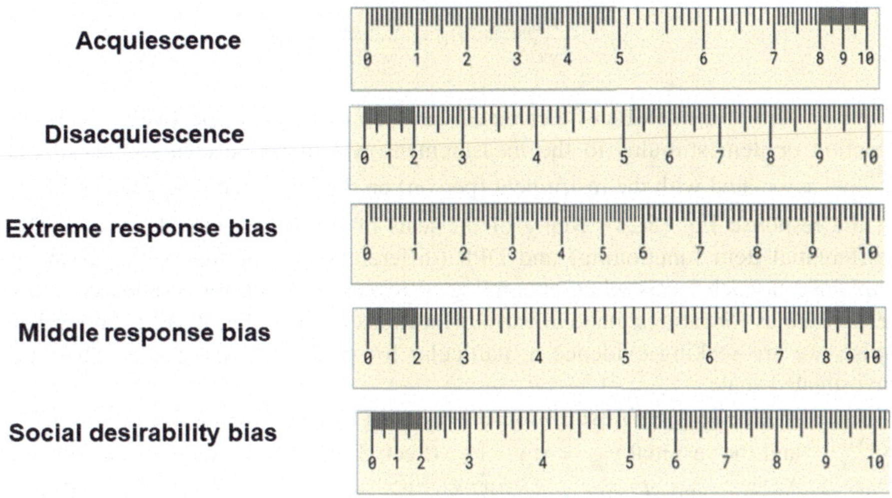

Fig. 5.16 Common response strategies to questionnaires and surveys

- Extreme response bias: 'Use scale endpoints regardless of item content'.
- Middle response bias: 'Use scale midpoint regardless of item content'.
- Social desirability bias: 'Present oneself in a positive way, regardless of item content'.

The various tests of the basic Rasch model, $P_{\text{success}} = \frac{e^{\theta-\delta}}{1+e^{\theta-\delta}}$ Eq. (1.1) provide support in assessing both (1) the validity and reliability of the model for a particular set of observations (Andrich 1988) and (2) metrological invariance (Tay et al. 2015; Humphry 2011; Pendrill and Fisher Jr. 2015; Putnick and Bornstein 2016) where the latter concerns the psychometric equivalence of a construct across different observations involving groups and/or across time. A variety of analyses to detect expected types of misfit to the Rasch model have been developed over the years by the groups of Mislevy (1986), Dardick (2010), Dardick and Mislevy (2016) and Wright (1995), to name a few. Most of the well-established battery of consistency checks on the Rasch model, which of course have to be made if one is to infer reliably metrological characteristics, can be described in our MSA performance metric framework.

Tests are made to reveal effects which, in one way or another, indicate a degree of breakdown in the basic Rasch assumption of determining item attribute values independent of person (instrument) attribute values (and vice versa). To first order, the residual for the *i*th probe or *j*th item can be modelled as:

$$\chi_i + \frac{\partial P_{\text{success}}}{\partial \theta} \cdot \partial \theta + \frac{\partial^2 P_{\text{success}}}{\partial \theta^2} \cdot \theta$$

$$\chi_j - \frac{\partial P_{\text{success}}}{\partial \delta} \cdot \partial \delta - \frac{\partial^2 P_{\text{success}}}{\partial \delta^2} \cdot \delta$$

that is, a sum of the effects of a changed value $\partial\theta$ or $\partial\delta$ of, respectively, the probe reaction or item stimulus to the measurement system and a changed sensitivity $\frac{\partial^2 P_{\text{success}}}{\partial \delta^2}$ associated with the instrument (person) on the original chi-squared residuals in the response Eq. (5.24). Many of the tests of the Rasch model, such as DIF (differential item functioning) and DPF (differential person functioning), can be explained in such terms when attempting to reveal evidence for acquiescence and response bias for ratings made at various locations of the scale. Apart from random noise, we are seeking evidence in particular for systematic distortions across the investigated scale.

Common measures of misfit, which have a long history, are in terms of $X = \chi_j^2 = \sum_{i=1}^{N_{\text{TP}}} y_i^2$ and the residual $y_{i,j} = x_{i,j} - \mathbb{E}_{i,j}$ (Sect. 5.6.2). These statistics include the $Chi^2 = \frac{X - N_{\text{TP}}}{\sqrt{2 \cdot N_{\text{TP}}} \cdot \sigma}$ and $Fisher = \frac{\sqrt{2 \cdot N_{\text{TP}}} - \sqrt{2 \cdot N_{\text{TP}} - 1}}{\sigma}$ as well as the Wilson–Hilferty (1931) statistic:

$$\text{WH} = \frac{2 \cdot \left(\sqrt[3]{X} - \sqrt[3]{N_{\text{TP}} - \frac{2}{3}} \right)}{\sigma} \tag{5.26}$$

In all these expressions, $\sigma = \frac{\sqrt{2}}{3 \cdot \sqrt[6]{N_{\text{TP}} - \frac{2}{3}}}$. These various versions of fit residuals differ, for instance, in speed of convergence to Normality as well as have different weighting factors (Wright and Masters 1990; WINSTEPS®; Wiki Chi-squared). The Wilson–Hilferty statistic Eq. (5.26) has become a more or less standard choice in some contemporary programs, such as WINSTEPS®, following the treatises of Wright and Stone (1979) and of Linacre (2010). A standardised weighted residual $z_{i,j}$ is calculated as $z_{i,j} = \frac{y_{i,j}}{\sqrt{W_{i,j}}}$, where the variance of the response x is $W_{i,j} = \sum_{k=0}^{C_j} \left(k - \mathbb{E}_{i,j}\right)^2 \cdot q_{i,j,k}$. For example, the weighted mean square INFIT MEANSQ:

$v_j = \frac{\sum_i^{N_{\text{TP}}} y_{i,j}^2}{\sum_i^{N_{\text{TP}}} W_{i,j}}$, while INFIT Zstd is $WH_{w,j} = \frac{2 \cdot \left(\sqrt[3]{v_j} - \sqrt[3]{N_{\text{TP}} - \frac{2}{3}} \right)}{\sigma}$. The regression residual INFIT Zstd is described in the literature as a kind of 'standardised *t*-factor'. Making the choice of Zstd, as opposed to other fit metrics such as MEANSQ (in WINSTEPS, for example), has been discussed by, for instance, Molton [http://www.rasch.org/poly.xls] and Smith et al. (2008). To quote Molton: 'Both statistics … require a lot of meditation and incense. For useful explanations, refer to: http://www.rasch.org/rmt/rmt162f.htm'.

A regular practice when employing the Rasch model is to check how well the data fit the model by plotting values (in a so-called Construct Alley) of a Rasch attribute parameter (either θ or δ) against the residuals, such as INFIT-ZSTD, of the logistic

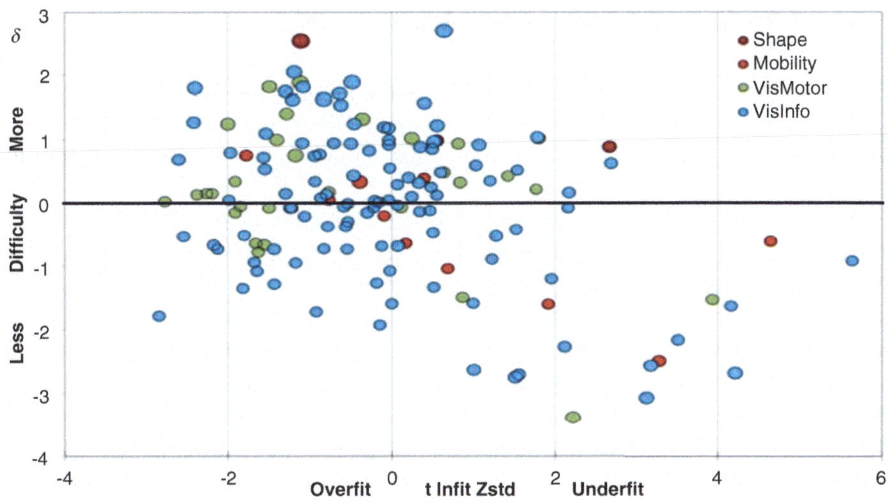

Fig. 5.17 Example of Infit zstd scores Eq. (5.26) for low-vision tasks (Reproduced with permission Jeter et al. 2017)

regression. Construct alleys such as shown in Fig. 5.17 can have a variety of fit residual factors plotted on the horizontal X-axis.

Typically one eliminates out of hand data points where the INFIT ZSTD measure Eq. (5.26) exceeds 2.5 σ. For instance, in a recent study of 232 sufferers of Myotonic Dystrophy Type 1, as many as 32 of 105 items of the DMI-Aktiv^{C} instrument were removed by Hermans et al. (2015) prior to a final Rasch analysis, where many data points had large (up 8 σ) deviations from the basic Rasch model. One has to ask whether it is reasonable to reject 30% of items without asking what valuable information might be lost by such a drastic elimination of 'outliers'? When treating outliers when employing the Rasch paradigm, statistical criteria are not by themselves usually considered sufficient grounds for rejection. The present work takes the approach, instead of rejecting outliers out of hand, of examining further possible reasons for these outliers for certain sets of data, and attempts to explain characteristic patterns of outliers in construct alley plots which have been reported but not to the best of our knowledge fully interpreted in the literature to date.

Massof (2014), Jeter et al. (2017) have reported systematic distortions in construct alleys for different vision traits (for tasks of mobility, reading, visual information and visual motor). As shown in Fig. 5.17, vision tasks such as reading a menu (Visinfo) fit the Rasch model with a broad range of INFIT Zstd scores Eq. (5.26) are distributed in construct alley plots differently from tasks such as cutting up food with a knife and fork (VisMotor). There is a general tendency for points to occupy the top-left and bottom-right quadrants of the construct alley plot; that is, scatter is roughly along a diagonal from large negative values of INFIT Zstd scores for the more difficult tasks (top-left) to large positive values for the easier tasks (bottom-right). Can that be explained?

5.7.1 Changes of Entity Scale: Acquiescence

Rescaling associated with the item stimulus at the input to the measurement system could be one plausible explanation for construct alleys, exemplified in Fig. 5.17, which show some degree of structure rather than a random, uncorrelated scatter of fit residuals.

As part of the debate on the validity of the Rasch model, Goldstein (2012) writes: 'the authors…claim that a "fundamental requirement" for measurement is that for every possible individual the "difficulty" order of all items is the same. This is ... extremely restrictive. ... I also find it difficult to see any theoretical justification for such invariance to be a desirable property of a measuring instrument'. In response, Linacre and Fisher (2012) write: 'Perhaps HG terms invariance restrictive because he misconceives it in some kind of absolute way, as Guttman (1977) did. In actual practice, the uncertainty ranges within which items fall vary across different kinds of applications. Screening tolerates more uncertainty than accountability, which tolerates more than diagnosis, and which can in turn tolerate more than research investigations of very small effect sizes'.

Mathematically, we treat stimulus rescaling with two steps: (A) modifying the scale of the Rasch parameter, for instance, δ, across a range of attribute values and (B) thereafter deriving the corresponding modified response overall of the measurement system expressed in terms of the INFIT Zstd metric (Eq. 5.26), bearing in mind the special sensitivity (Eq. (5.16), and Fig. 5.15) of a measurement system with Man as a measurement instrument.

When modelling at step (A) the effects of a change of stimulus on the response, there are, of course, different conceivable models for the change in Rasch parameter such as might be associated with effects such as acquiescence and response bias at various locations of the scale. A simple rescaling, centred and varying linearly would be $\partial\delta_j = \tau \cdot \delta_j$, where τ is a rescaling factor. For some reason or other (Fig. 5.16), rating of item j is made so that the item attribute (such as task difficulty) lies on a different scale: a positive value τ of rescaling indicates an extended scale [test persons (instruments) rate this item more strongly than others, perhaps to indicate an increased importance or weight], while a negative value τ of rescaling corresponds to the case where rating does not recognise a reversed scale. An example of the latter is where a survey designer has deliberately included alternately positive (true key) and negative (false key) items, perhaps to reveal evidence of acquiescence in raters, that is, responses which tend to agree with questions (e.g. personality scales (Soto et al. 2008)) without due regard for the content of the item.

The resulting modified response can be obtained by substituting the weighted residual expression in the various goodness-of-fit statistics, where $X = \chi_j^2 = \sum_{i=1}^{N_{TP}} y_i^2$(Eq. 5.24) and the residual $y_{i,j} = x_{i,j} - \mathbb{E}_{i,j}$ (Sect. 5.6.2).

Responses modified by a change in stimulus, such as a *rescaling*, $\partial\theta_i = \tau \cdot \theta_i$ or $\partial\delta_j = \tau \cdot \delta_j$ lead, respectively, to modified chi-squared statistics given by

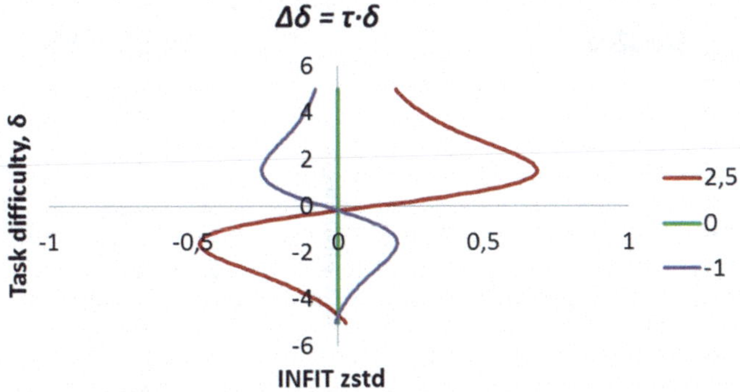

Fig. 5.18 Construct alley plots of Rasch δ attribute value over a range of items against the fit statistic INFIT zstd, Wilson–Hilferty (WH) Eq. (5.26), for different rescalings τ $\partial\boldsymbol{\delta}_j = \boldsymbol{\tau} \cdot \boldsymbol{\delta}_j$ Eq. (5.27)

$$\chi'_{i,\theta}{}^2 = \left(\chi_i + \frac{\partial P_{\text{success}}}{\partial\theta} \cdot \partial\theta\right)^2 = \chi_i^2 + 2 \cdot \frac{\partial P_{\text{success}}}{\partial\theta} \cdot \partial\theta \cdot \chi_i + \left(\frac{\partial P_{\text{success}}}{\partial\theta} \cdot \partial\theta\right)^2$$
$$\chi'_{j,\delta}{}^2 = \left(\chi_j - \frac{\partial P_{\text{success}}}{\partial\delta} \cdot \partial\delta\right)^2 = \chi_j^2 - 2 \cdot \frac{\partial P_{\text{success}}}{\partial\delta} \cdot \partial\delta \cdot \chi_j + \left(\frac{\partial P_{\text{success}}}{\partial\delta} \cdot \partial\delta\right)^2 \quad (5.27)$$

At step (B), the special 'resonance-like' behaviour of the measurement system sensitivity $K = \frac{\partial P_{\text{success}}}{\partial\delta}$ (shown in Fig. 5.15) is expected to have radical effects on the regression residuals in the response of the measurement system.

Typical results of such a rescaling analysis are shown in Fig. 5.18, with the distinctive, 'dispersion-like' shape of each construct alley plot of Wilson–Hilferty statistic WH evaluated using Eqs. (5.26) and (5.27). Corresponding plots for the other goodness-of-fit statistics—chi-squared and Fisher—yield substantially similar results and we found for typical data that differences between the Fisherian and Wilson–Hilferty statistics were negligibly small compared with the effects of rescaling the residuals.

As expected, as shown in Figs. 5.18 and 5.19, at either extreme of the item attribute scale the effects of rescaling converge to zero, while the largest displacements are found close to midscale ($\delta \sim 0$; $P_{\text{success}} \sim 50\ \%$): this reflects the 'resonance-like' variation in sensitivity (Fig. 5.15) where the Rasch model is largely insensitive to large excursions at the scale extremes. Depending on the degree of scale distortion, compared with the average, responses to items which are judged on broader (more generous) scales ($\Delta\delta = +2.5 \cdot \delta$) lead to greater excursions in the top-right and bottom-left quadrants of the construct alley plots than the narrower (more conservative) scale ($\Delta\delta = 0$) (Fig. 5.19a). Acquiescence (e.g. $\tau = -1$) is distinguished

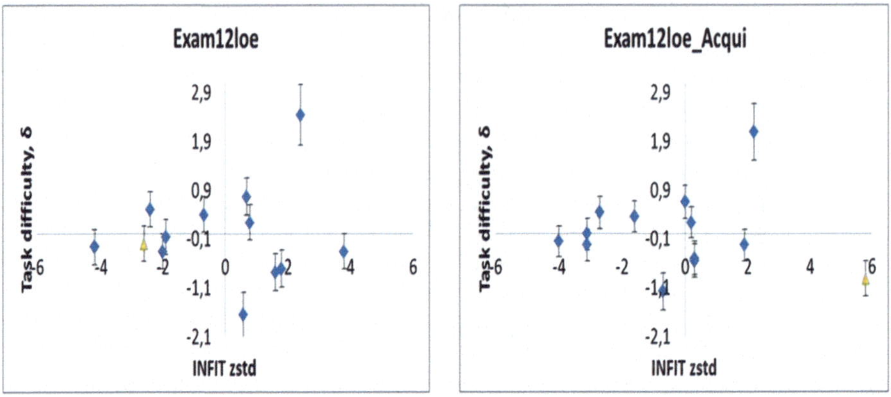

k = 2

Fig. 5.19 Construct alley plots of Rasch δ attribute value over a range of items against the fit statistic INFIT zstd, Wilson–Hilferty (WH) Eq. (5.26), for different rescaling $\tau\ \partial\delta_j = \tau \cdot \delta_j$ for the WINSTEPS® example EXAM12 (a) original FIM™ data; (b) simulated acquiescence for one item (yellow triangle; toilet transfer) where scale reversed ($\tau = -1$) for whole cohort of 35 arthritis patients

correspondingly by construct alleys which fill the top-left/bottom-right quadrants of plots such as Fig. 5.19b.

The effects of rescaling due to acquiescence can be easily simulated, taking, for example, the data set 'EXAM12' provided with the WINSTEPS® program, described as follows: '35 arthritis patients have been through rehabilitation therapy. Their admission to therapy and discharge from therapy measures are to be compared. They have been rated on the 13 mobility items of the functional independence measure (FIM™). Each item has seven levels ((1) 0% Independent; (2) 25% Independent; (3) 50% Independent; (4) 75% Independent; (5) Supervision; (6) Device; (7) Independent). At admission, the patients could not perform at the higher levels. At discharge, all patients had surpassed the lower levels (Data courtesy of C.V. Granger & B. Hamilton, ADS). The admission ratings are in EXAM12LO. TXT'.

5.7.2 *Changes of Sensitivity*

A similar procedure, with steps (A) and (B), can be followed to describe the change in response following a *change in sensitivity* of the instrument (person) at the heart of the measurement system, expressed as a modified chi-squared statistic:

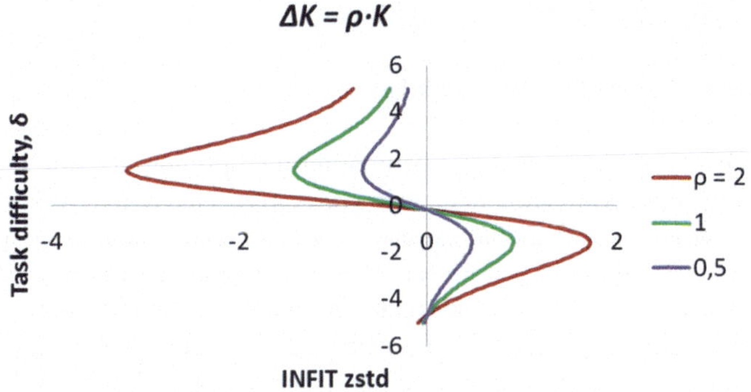

Fig. 5.20 Construct alley plots of Rasch δ attribute value over a range of items against the fit statistic INFIT zstd, Wilson–Hilferty (WH) Eq. (5.26), for different discrimination values ρ $\frac{\partial^2 P_{\text{success}}}{\partial \delta^2} = \rho \cdot \frac{\partial P_{\text{success}}}{\partial \delta}$ Eq. (5.28)

$$\left(\chi_j - \frac{\partial^2 P_{\text{success}}}{\partial \delta^2} \cdot \delta\right)^2 = \chi_j^2 - 2 \cdot \frac{\partial^2 P_{\text{success}}}{\partial \delta^2} \cdot \delta \cdot \chi_j + \left(\frac{\partial^2 P_{\text{success}}}{\partial \delta^2} \cdot \delta\right)^2 \qquad (5.28)$$

where $X = \chi_j^2 = \sum_{i=1}^{N_{\text{TP}}} y_i^2$and the residual $y_{i,j} = x_{i,j} - \mathbb{E}_{i,j}$ (Sect. 5.6.2).

Such a change in sensitivity $\frac{\partial^2 P_{\text{success}}}{\partial \delta^2} = \rho \cdot \frac{\partial P_{\text{success}}}{\partial \delta}$ could be described in terms of the discrimination factor ρ of the 2PL IRT model (Partchev 2004).

In fact we can independently calculate

$$\text{INFIT\_zstd} = \frac{\partial K}{\partial \delta} = \frac{\partial^2 P_{\text{success}}}{\partial \delta^2} = \frac{e^{(\theta-\delta)}}{1 + e^{(\theta-\delta)}} - \frac{3 \cdot e^{2 \cdot (\theta-\delta)}}{\left(1 + e^{(\theta-\delta)}\right)^2} + \frac{2 \cdot e^{(\theta-\delta)} \cdot e^{2 \cdot (\theta-\delta)}}{\left(1 + e^{(\theta-\delta)}\right)^3}$$

As hypothesised by Humphry (2011), distinct values of the empirical factor ρ_s for each set, s, of classifications in the logistic measurement function Eq. (3.7) can affect the slope of the item response function. The change in slope of the item response function has its maximum at the origin, while slope changes converge to zero at either extreme of the logit scale. For a certain difference Δz (arising, for example, from a change in scale), the corresponding change $\Delta P_{\text{success}}$ will vary greatly depending on where z lies, since the sensitivity of the instrument is strongly non-linear, as is evident from Figs. 5.15 and 5.20. The greatest change $\Delta P_{\text{success}}$ and the change of sign of $\Delta P_{\text{success}}$ from plus to minus (or vice versa) will occur of course near the $P_{\text{success}} = 50\%$ mid-range where the sensitivity has its maximum, while at either extreme end of the scale, $\Delta P_{\text{success}}$ will tend asymptotically to zero.

5.7.3 Further Tests of Assumptions

Logistic Regression and Estimator Bias

There a number of different ways of estimating the optimum logistic regression of the Rasch model to the raw score data, using principles such as maximum likelihood, weighted likelihood, Bayes modal estimation and so on, as examined, for instance, by Lord (1983), Hoijtink and Boomsma (1996) and Linacre (2004a, b, c).

It is well-known that there will be some bias and variance in the parameter values of the response variable when estimated with statistical inference in general (not just of the Rasch model). The different principles of inference will have different biases.

For the jth item, $P_{\text{success},\ j} = P(X_j = x \mid \theta, \delta_j)$ describes the PDF of the item response X_j conditional on a corresponding distribution of person ability θ and for a given item difficulty δ_j. The maximum likelihood estimate (MLE) of θ is obtained by solving: $\frac{\partial \ln \boldsymbol{P}(\boldsymbol{U} = \boldsymbol{u}|\boldsymbol{\theta}\ ,\ \boldsymbol{\delta})}{\partial \theta} = 0,$ where $\boldsymbol{P}(\boldsymbol{U} = \boldsymbol{u}|\boldsymbol{\theta}\ ,\ \boldsymbol{\delta}) = \prod_j P_{\text{success},j} \cdot \left[1 - P_{\text{success},j}\right]$ for the Rasch model, assuming local stochastic independence among the items, j. According to Hoijtink and Boomsma (1996), the (log)likelihood is stationary at a value of θ, where $\sum_j X_j = \sum_j P_{\text{success},j}$.

According to Lord (1983), there is a bias of the resulting maximum likelihood estimate of θ:

$$\text{Bias}(\text{MLE}(\theta)) = \mathbb{E}(\text{MLE}(\theta)) - \theta = \frac{-J(\text{MLE}(\theta))}{2 \cdot I(\text{MLE}(\theta))^2} \tag{5.29}$$

where for the Rasch model: $J[\text{MLE}(\theta)] = \sum_j \frac{\frac{\partial^2 P_{\text{success},j}}{\partial^2 \theta_j} \cdot \frac{\partial P_{\text{success},j}}{\partial \theta}}{P_{\text{success},j} \cdot \left(1 - P_{\text{success},j}\right)}$.

The covariance of the maximum likelihood estimate is the inverse of the Fisher information: $\text{Var}(\text{MLE}(\theta)) = \frac{1}{I(\text{MLE}(\theta))}$ (Rao 1973; Agresti 2013). The Fisher information $I(\theta)_{i,j} = -\mathbb{E}\left[\frac{\partial^2 L(\theta)}{\partial \theta_i \cdot \partial \theta_j}\right]$ with the log-likelihood[4] for the Rasch model: $L(\theta) = \ln\left[l(\theta)\right] = \ln\left(\frac{P_{\text{success},j}}{1 - P_{\text{success},j}}\right) = \theta - \delta_j$ for the jth item. In that case, the Fisher information is $\left[(\theta)\right] = \sum_j \frac{\frac{\partial P_{\text{success},j}}{\partial \theta}^2}{P_{\text{success},j} \cdot \left(1 - P_{\text{success},j}\right)}$.

Estimating Rasch parameters by maximising the joint log-likelihood does not provide consistent estimates according to (Christensen and Olsbjerg 2013), since the number of parameters increases with sample size, echoing the 'nuisance' parameters.

[4]It is easier to maximise the log-likelihood (which has the same maximum as the likelihood) since it is a sum rather than a product of terms, as in the case of the Rasch model.

Despite these reservations about the accuracy of statistical inference of parameters in the Rasch model, in practice the biases are expected in many cases to be small compared with other imprecisions. Linacre in his documentation of the Winsteps® program notes:

> JMLE exhibits some estimation bias in small data sets, but this rarely exceeds the precision (model standard error of measurement, SEM) of the measures. Estimation bias is only of concern when exact probabilistic inferences are to be made from short tests or small samples. It can be exactly corrected for paired-comparison data with PAIRED=Yes. For other data, it can be approximately corrected with STBIAS=Yes, but, in practice, this is not necessary (and sometimes not advisable).

References

Y Akao, G H Mazur, The leading edge in QFD: past, present and future, Int. J. Quality & Reliability Management **20**, 20–35 (2003)

A. Agresti, *Categorical Data Analysis*, 3rd edn. (Wiley, Hoboken, 2013). ISBN 978-0-470-46363-5

D. Andrich, *Rasch Models for Measurement* (Sage Publications, Beverly Hills, 1988)

D. Andrich, A law of ordinal random error: The Rasch measurement model and random error distributions of ordinal assessments. J. Phys. Conf. Series **1044**, 012055 (2017). https://doi.org/10.1088/1742-6596/1044/1/012055. J. of Phys. Conference Series IMEKO 2017, IOP Conf. Series.

F. Attneave, Informational aspects of visual perception. Psychol. Rev. **61**, 183–193 (1954)

H. Barlow, The exploitation of regularities in the environment by the brain. Behav. Brain Sci. **24**, 602–607 (2001)

E. Bashkansky, S. Dror, Matrix approach to analysis of human errors and their prevention by quality engineering and managerial tools. Qual. Reliab. Eng. Int. *32*(2), 535–545 (2015). https://doi.org/10.1002/qre.1770

E. Bashkansky, T. Gadrich, Some metrological aspects of ordinal measurements. Accredit. Qual. Assur. **15**, 331–336 (2010). https://doi.org/10.1007/s00769-009-0620-x

M.J. Bitner, B.H. Booms, M. Stanfield Trereault, The service encounter: diagnosing favorable and unfavorable incidents. J. Mark. **54**, 71–84 (1990)

J.P. Bentley (2005) Principles of Measurement Systems, Pearson Education Limited www.pearsoned.co.uk 4th edition ISBN 0 130 43028 5

M. Boers, J.R. Kirwan, P. Tugwell, et al., *The OMERACT Handbook* (OMERACT, Ottawa, 2018). https://omeract.org/resources

L. Brillouin, Science and information theory, in *Physics Today*, vol. 15, 2nd edn., (Academic Press, Cambridge, 1962). https://doi.org/10.1063/1.3057866

L. Carnot, *Principes Fondamentaux de l'Équilibre et du Movement* [Fundamental Principles of Equilibrium and Movement], Paris (1803)

L K. Chan, M.L. Wu, Quality function deployment: a literature review, Eur. J. Operational Research **143**, 463–97 (2002), https://doi.org/10.1080/00207540600575779

L. Chen, X. Liang, T. Li, Collaborative performance research on multi-level hospital management based on synergy entropy-HoQ. Entropy **17**, 2409–2431 (2015). https://doi.org/10.3390/e1704240

B. Christe, E. Cooney, R. Rogers, Analysis of the impact of a radiofrequency identification asset-tracking system in the healthcare setting. J. Clin. Eng. **35**, 49–55 (2010)

K.B. Christensen, M. Olsberg, Marginal maximum likelihood estimation in polytomous Rasch models using SAS, in *Annales de l'ISUP* (2013), https://www.researchgate.net/publication/

258396258_Conditional_Maximum_Likelihood_Estimation_in_Polytomous_Rasch_Models_Using_SAS. Accessed 31 Aug 2018

L. J. Cronbach, R. Nageswari, and G.C. Gleser (1963). Theory of generalizability: A liberation of reliability theory. The British Journal of Statistical Psychology, 16, 137–163

J. Dagman, R. Emardson, S. Kanerva, L.R. Pendrill, A. Farbrot, S. Abbas, A. Nihlstrand, Measuring comfort for heavily-incontinent patients assisted by Absorbent products in several contexts, in *Incontinence in The Engineering Challenge IX*, IMECHE, London, 5–6 November 2013 (2013)

W.R. Dardick, *Reweighting Data in the Spirit of Tukey: Using Bayesian Posterior Probabilities as Rasch Residuals for Studying Misfit*, Dissertation, University of Maryland (2010)

W.R. Dardick, R.J. Mislevy, Reweighting data in the spirit of Tukey. Educ. Psychol. Meas. **76**, 88–113 (2016). https://doi.org/10.1177/0013164415583351

J.G. Dolan, Shared decision-making – Transferring research into practice: the analytic hierarchy process (AHP). Patient Educ. Couns. **73**, 418–425 (2008). https://doi.org/10.1016/j.pec.2008.07.032

N. Dorans, T. Moses, D. Eignor, *Principles and Practices of Test Score Equating* (Educational Testing Service, Princeton, 2010). ETS RR-10-29.

G.H. Fischer, The linear logistic test model as an instrument in educational research. Acta Psychol. **37**, 359–374 (1973). https://www.sciencedirect.com/science/article/pii/0001691873900036

W. P. Fisher Jr. (2015) Rasch measurement as a basis for metrologically traceable standards Rasch Meas. Trans. **28** 1492–3

J.A. Fisher, T. Monahan, Evaluation of real-time location systems in their hospital contexts. Int. J. Med. Inform. **81**, 705–712 (2012)

R. Fleischmann, Einheiteninvariante Gröβengleichungen, Dimension. Der Mathematische und Naturwissenschaftliche Unterricht **12**, 386–399 (1960)

T. Gadrich, E Bashkansky, Ricardas Zitikis (2014), Assessing variation: a unifying approach for all scales of measurement, Qual. Quant., http://dx.doi.org/10.1007/s11135-014-0040-9

H. Goldstein, Francis Galton, measurement, psychometrics and social progress. Assess. Educ. Princ Policy Pract. **19**(2), 147–158 (2012). www.bristol.ac.uk/cmm/team/hg/full-publications/2012/Galton.pdf

D.D. Gremler, The critical incident technique in service research. J. Serv. Res. **7**, 65–89 (2004)

L. Guttman, What is not what in statistics. Stat. **26**, 81–107 (1977)

Y. Heerkens, J. Engels, C. Kuiper, J. Van Der Gulden, R. Oostendorp, The use of the ICF to describe work related factors influencing the health of employees. Disabil. Rehabil. **26**, 1060–1066 (2004). https://doi.org/10.1080/09638280410001703530

J.C. Helton, Uncertainty and sensitivity analysis in the presence of stochastic and subjective uncertainty. J. Stat. Comput. Simul. **57**, 3–76 (1997)

J.C. Helton et al., Survey of sampling-based methods for uncertainty and sensitivity analysis. Reliab. Eng. Syst. Saf. **91**, 1175–1209 (2006)

M.C.E. Hermans, J.G.J. Hoeijmakers, C.G. Faber, I.S.J. Merkies, Reconstructing the Rasch-built Myotonic dystrophy type 1 activity and participation scale. PLoS One **10**(10), e0139944 (2015). https://doi.org/10.1371/journal.pone.0139944

H. Hoijtink, A. Boomsma, Statistical inferences based on latent ability estimates. Psychometrika **61**, 313–330 (1996)

S.M. Humphry, The role of the unit in physics and psychometrics. Measurement **9**, 1–24 (2011)

JCGM 100:2008 Evaluation of measurement data – Guide to the expression of uncertainty in measurement (GUM 1995 with minor corrections) in Joint Committee on Guides in Metrology (JCGM)(2008)

P. Jeter, C. Rozanski, R. Massof, O. Adeyemo, G. Dagnelie, the PLOVR Study Group, Development of the ultra-low vision functioning questionnaire (ULV-VFQ). Transl. Vis. Sci. Technol. **6** (3), 11 (2017). https://doi.org/10.1167/tvst.6.3.11

W. Kintsch, *The Representation of Meaning in Memory* (Lawrence Erlbaum Associates, Hillsdale, 1974)

G.J. Klir, T.A. Folger, *Fuzzy Sets, Uncertainty and Information* (Prentice Hall, Upper Saddle River, 1988). ISBN 0-13-345984-5

H. Knox, A scale, based on the work at Ellis Island, for estimating mental defect. J. Am. Med. Assoc. **LXII**(10), 741–747 (1914)

S. Kullback, R.A. Leibler, On information and sufficiency. Ann. Math. Statist. **22**, 79–86 (1951). https://doi.org/10.1214/aoms/1177729694

S.L. Latimer, Using the linear logistic test model to investigate a discourse-based model of Reading comprehension. Educ. Res. Perspect. **9**(1), 73–94 (1982). https://www.rasch.org/erp6.htm

J.M. Linacre, The Rasch model cannot be "disproved"! Rasch Meas. Trans. **10**(3), 512–514 (1996)

J.M. Linacre, Optimizing rating scale category effectiveness. J. Appl. Meas. **3**(1), 85–106 (2002)

J.M. Linacre, "Estimation methods for Rasch measures", Chapter 2, in *Introduction to Rasch Measurement,* ed. by E.V. Smith, R.M. Smith (JAM Press, Maple Grove, 2004a)

J.M. Linacre, "Rasch model estimation: further topics", Chapter 24, in *Introduction to Rasch Measurement,* ed. by E.V. Smith, R.M. Smith (JAM Press, Maple Grove, 2004b).

J.M. Linacre, Discrimination, Guessing and carelessness asymptotes: estimating IRT parameters with Rasch. Rasch Meas. Trans. **18**(1), 959–960 (2004c)

J.M. Linacre, Rasch model with an error term. Rasch Meas. Trans. **23**, 1238 (2010)

J.M. Linacre, W.P. Fisher Jr., Harvey Goldstein's objections to Rasch measurement: a response from Linacre and Fisher. Rasch Meas. Trans. **26**(3), 1383–1389 (2012)

J.M. Linacre, J.W. Heinemann, B.D. Wright, C.V. Granger, B.B. Hamilton, The structure and stability of the functional independence measure. Arch. Phys. Med. Rehabil. **75**, 127–132 (1994)

F.M. Lord, Unbiased estimators of ability parameters, of their variance, and of their parallel-forms reliability. Psychometrika **48**, 233–245 (1983)

R.W. Massof, Are subscales compatible with univariate measures, in *IOMW 2014 workshop, Philadelphia (PA, USA)* (2014). https://sites.google.com/site/iomw2014archive/iomw-program

R.W. Massof, C. Bradley, A strategy for measuring patient preferences to incorporate in benefit-risk assessment if new ophthalmic devices and procedures. J Phys Conf Ser. **772**, 012047 (2016). https://doi.org/10.1088/1742-6596/772/1/012047

R.W. Massof, L. Ahmadian, L.L. Grover, J.T. Deremeik, J.E. Goldstein, C. Rainer, C. Epstein, G.D. Barnett, The activity inventory: an adaptive visual function questionnaire. Optom. Vis. Sci. **84**, 763–774 (2007)

J. Melin, S.J Cano, L.R Pendrill, Metrology of human-based perceptions: The role of entropy in construct specification equations to improve the validity of cognitive tests, in *Perspectives on Science (*POS*),* Measurement at the Crossroads special issue, submitted 190520 (2019)

A. Mencattini, L. Mari, A conceptual framework for concept definition in measurement: The case of 'sensitivity'. Measurement **72**, 77–87 (2015)

G.A. Miller, The magical number seven, plus or minus two: some limits on our capacity for processing information. Psychol. Rev. **63**, 81–97 (1956)

R.J. Mislevy, Bayes model estimates in item response models. Psychometrika **51**, 177–195 (1986)

G. Nordin, R. Dybkaer, U. Forsum, X. Fuentes-Arderiu, F. Pontet, Vocabulary on nominal property, examination, and related concepts for clinical laboratory sciences (IFCC-IUPAC recommendations 2017). Pure Appl. Chem. **90**(5), 913–935 (2018). https://doi.org/10.1515/pac-2011-0613

I. Partchev, *A Visual Guide to Item Response Theory* (Friedrich-Schiller-Universität Jena, Jena, 2004). https://www.coursehero.com/file/28232270/Partchev-VisualIRTpdf/

O. Pele, M. Werman, The quadratic-chi histogram distance family, in *Computer Vision–ECCV 2010* (2010). pp. 749–762. www.cs.huji.ac.il/~werman/Papers/ECCV2010.pdf

L.R. Pendrill, Man as a measurement instrument. *NCSLI Meas. J. Meas. Sci.* **9**, 24–35 (2014a)

L.R. Pendrill, Using measurement uncertainty in decision-making & conformity assessment. Metrologia **51**, S206 (2014b). https://doi.org/10.1088/0026-1394/51/4/S206

L.R. Pendrill, Limits to the reliability of the Rasch psychometric model. ENBIS 2017 (Naples, Sept.) (2017)

L.R. Pendrill, W.P. Fisher Jr., Counting and quantification: comparing psychometric and metrological perspectives on visual perceptions of number. Measurement **71**, 46–55 (2015). https://doi.org/10.1016/j.measurement.2015.04.010

L.R. Pendrill, J. Melin, S. Cano and the NeuroMET consortium 2019, Metrological references for health care based on entropy, 19th International Congress of Metrology, Paris (FR), EDP Science: web of conference open access, https://cfmetrologie.edpsciences.org/component/issues/, in press

L.R. Pendrill, N. Petersson, Metrology of human-based and other categorical measurements. Meas. Sci. Technol. **27**, 094003 (2016). https://doi.org/10.1088/0957-0233/27/9/094003

A. Possolo, Measurement, in *Proceedings AMCTM 2017* (2018)

A. Possolo, C. Elster, Evaluating the uncertainty of input quantities in measurement models. Metrologia **51**, 339–353 (2014). https://doi.org/10.1088/0026-1394/51/3/339

D.L. Putnick, M.H. Bornstein, Measurement invariance conventions and reporting: the state of the art and future directions for psychological research. Dev. Rev. **41**, 71–90 (2016)

C.R. Rao, *Linear Statistical Inference and its Applications*, 2nd edn. (Wiley, Hoboken, 1973)

G. Rasch, On general laws and the meaning of measurement in psychology, in *Proceedings of the Fourth Berkeley Symposium on Mathematical Statistics and Probability*, vol. IV, (University of California Press, Berkeley, 1961), pp. 321–334. Available free from Project Euclid

K.E. Roach, Measurement of health outcomes: Reliability, validity and responsiveness. J. Prosthet Orthot **18**, 8 (2006)

G.B. Rossi, Measurement and probability – A probabilistic theory of measurement with applications, in *Springer Series in Measurement Science and Technology* (2014). https://doi.org/10.1007/978-94-017-8825-0

Y. Rubner, C. Tomasi, L.J. Guibas, The earth mover's distance as a metric for image retrieval. Int. J. Comput. Vis. **40**, 99–121 (2000)

H.H. Scheiblechner, CML-parameter-estimation in a generalized multifactorial version of Rasch's probabilistic measurement model with two categories of answers, in *Research Bulletin*, vol. 4, (Psychologisches Institut det Universität Wien, Vienna, 1971)

T.D. Schneider, G.D. Stormo, L. Gold, A. Ehrenfeuch, The information content of binding sites on nucleotide sequences. J. Mol. Biol. **188**, 415–431 (1986). www.lecb.ncifcrf.gov/~toms/paper/schneider1986

M.M. Schnore, J.T. Partington, Immediate memory for visual patterns: symmetry and amount of information. Psychon. Sci. **8**, 421–422 (1967)

A.B. Smith, R. Rush, L.J. Fallowfield, G. Velikova, M. Sharpe. Rasch fit statistics and sample size considerations for polytomous data, BMC Medical Research Methodology, **8**, 1–11, https://doi.org/10.1186/1147-2288-8-33 (2008)

C.J. Soto, O.P. John, S.D. Gosling, J. Potter, The developmental psychometrics of big five self-reports: Aacquiescence, factor structure, coherence, and differentiation from ages 10 to 20, J. Pers. Soc. Psychol., **94**, 714–37 (2008)

A.J. Stenner, M. Smith, Testing construct theories. Percept. Mot. Skills **55**, 415–426 (1982). https://doi.org/10.2466/pms.1982.55.2.415

A.J. Stenner, I.I.I. M Smith, D.S. Burdick, Toward a theory of construct definition. J. Educ. Meas. **20**(4), 305–316 (1983)

A.J. Stenner, W.P. Fisher Jr., M.H. Stone, D.S. Burdick, Causal Rasch models. Front. Psychol. **4** (536), 1–14 (2013)

E. Svensson, Guidelines to statistical evaluation of data from rating scales and questionnaires. J. Rehabil. Med. **33**, 47–48 (2001)

L. Tay, A.W. Meade, M. Cao, An overview and practical guide to IRT measurement equivalence analysis. Organ. Res. Methods **18**, 3–46 (2015)

J.A. Tukey, Chapter 8, Data analysis and behavioural science, in The Collected Works of John A Tukey, Volume III, Philosophy and Principles of Data Analysis: 1949–1964, ed. by L. V. Jones, (University North Carolina, Chapel Hill, 1986)

A.M. van der Bles, S. van der Linden, A.L.J. Freeman, J. Mitchell, A.B. Galvao, L. Zaval, D.J. Speigelhalter, Communicating uncertainty about facts, numbers and science. R. Soc. Open Sci. **6**, 181870 (2019). https://doi.org/10.1098/rsos.181870

A. Verhoef, G. Huljberts and W. Vaessen, (2015), Introduction of a quality index, based on Generalizablilty theory, as a measure of rellability for univariate- and multivariate sensory descriptive data. Food quality and Preference, **40**, 296–303

W. Weaver, C. Shannon, *The Mathematical Theory of Communication* (Univ. of Illinois Press, Champaign, 1963). ISBN 0252725484

Wiki Chi-squared distribution. https://en.wikipedia.org/wiki/Chi-squared_distribution

E.B. Wilson, M.M. Hilferty, The distribution of chi-square. Proc. NAS **17**, 684–688 (1931)

WINSTEPS® manual, p. 284. http://www.winsteps.com/index.htm

S. Wold et al. PLS-regression: a basic tool of chemometrics. Chemometrics and Intelligent Laboratory Systems **58**, 109–13 (2001)

B.D. Wright, Comparing factor analysis and Rasch measurement. Rasch Meas. Trans. **8**(1), 350 (1994). http://www.rasch.org/rmt/rmt81r.htm

B.D. Wright, Diagnosing person misfit. Rasch Meas. Trans. **9**(2), 430–431 (1995)

B.D. Wright, G.N. Masters, Computation of OUTFIT and INFIT statistics. Rasch Meas. Trans. **3** (4), 84–85 (1990). http://www.rasch.org/rmt/rmt34e.htm

B.D. Wright, M.H. Stone, *Best Test Design* (MESA Press, Chicago, 1979). ISBN 0-941938-00-X. LC# 79-88489.

J. Yang, W. Qiu, A measure of risk and a decision-making model based on expected utility and entropy. Eur. J. Oper. Res. **164**, 792–799 (2005)

Y. Yao, W.L. Lu, B. Xu, C.B. Li, C.P. Lin, D. Waxman, J.F. Feng, The increase of the functional entropy of the human brain with age. Sci. Rep. **3**, 2853 (2013). https://doi.org/10.1038/srep02853. www.nature.com/scientificreports

J.V. Zidek, C. van Eeden, Uncertainty, Entropy, Variance and the Effect of Partial Information. Lect. Notes Monogr. Ser. **42**, 155–167 (2003). Mathematical Statistics and Applications: Festschrift for Constance van Eeden. https://projecteuclid.org/download/pdf_1/euclid.lnms/1215091936

Chapter 6
Decisions About Product

6.1 Use of Measurement Results and Conformity Assessment

Conformity assessment is making a decision about whether a product, service or other entity conforms to specifications. This final chapter deals with putting into use, when making decisions, all the measurement tools given in the intervening chapters to demonstrate to what extent actual measurement results live up to the initial motivations for making conformity assessment (providing consumer confidence; tools for the consumer and supplier when ensuring product quality; essential for several reasons, such as health, environmental protection, fair trade and so on) presented in Sect. 1.1.

Quality assured measurement across the social and physical sciences can be placed, as illustrated in Fig. 6.1, in a wider quality infrastructure providing a framework for equitable conformity assessment, linking requirements, regulatory guidelines, written standards and metrology (quality assured measurement). An ambition to achieve better products or services, for example, in care, while maintaining patient safety and personal data protection, can lead in different ways to the formation of regulatory requirements where, for example, specifications can be presented in a harmonised and transparent manner in written standards. Quality assured measurements should be made to check that the specifications set out in relevant standards are related to requirements. And finally, experiences about the advantages and disadvantages gained in a certain case of quality assurance can be fed back to the decision-makers in terms of best practice when embarking on a further loop from policy, regulation, standardisation and measurement.

Quality assurance of product is thus intimately related, as said previously, to the quality of measurement—comparability of product quality characteristics is obtained by measuring product with comparable measurement, as assured by metrological traceability to agreed and common reference standards. Measurement uncertainty leads to certain risks of incorrect decisions in conformity assessment.

L. Pendrill, *Quality Assured Measurement*, Springer Series in Measurement Science and Technology, https://doi.org/10.1007/978-3-030-28695-8_6

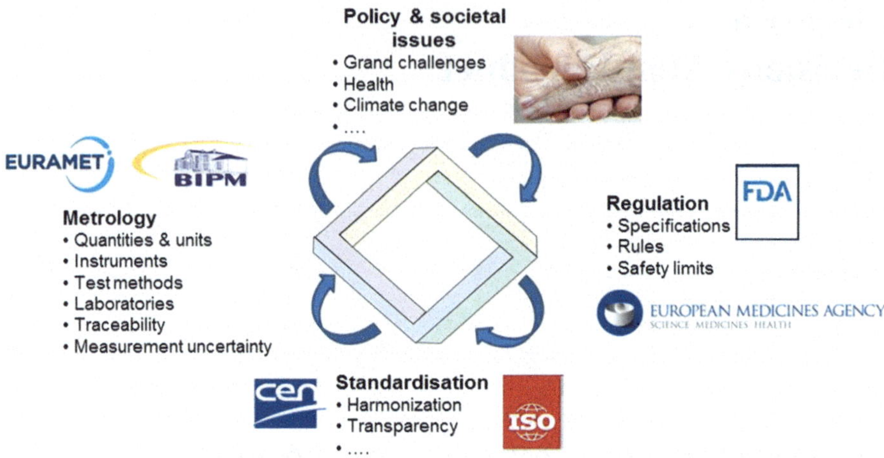

Fig. 6.1 Quality infrastructure: Linking requirements, regulatory guidelines, written standards and metrology

Design of experiment, that is, planning ahead so that resources expended on measurement will stand in reasonable proportion to the expected costs of not making perfect measurements, has been explained, where we have considered the relative importance of these two aspects—traceability and uncertainty—of metrology in the context of conformity assessment. Rules-of-thumb that uncertainties associated with lack of traceability might be reasonably limited to half of the total measurement uncertainty in any measurement result as a first working hypothesis. In this closing chapter, the predictions of these and more insightful judgements about 'fit-for-purpose' measurement based on cost and impact will be revisited with the actual measurement results in hand, such as obtained in the pre-packaged goods example followed throughout the book.

6.2 Closing the Quality Assurance Loop

The quality assurance loop opened in the first chapter will now be finally closed (to be opened at a later date when and if a decision is taken to develop a new, innovative product, learning from the strengths and weaknesses of previous product). When closing the loop, we so to say return to where we started, where final judgement is made of whether product actually satisfies the requirement stipulated or not and expectations met.

Descriptions of how test results are obtained (step (c) in the loop as given in Sect. 1.3) have been provided in Chaps. 2, 3, 4 and 5. Judgement about the entity will sometimes be confounded by measurement uncertainty.

A first effect of imperfect measurement for both consumer and supplier, due to limited sampling and finite measurement uncertainties, will be to possibly make incorrect estimates of entity error. This will include both examining apparent differences in response (Sect. 6.3) as well as differences in measurement 'value' even in the literal sense (Sect. 6.4).

Secondly, imperfect measurement will also increase the risks of making incorrect decisions of conformity to a specification limit, such as failing a conforming entity or passing a non-conforming entity when the test result is close to a tolerance limit (Sect. 6.4), as occur in the remaining steps of the quality loop:

(d) Decide if test results indicate that the entity, as well as the measurements themselves, is within specified requirements or not (Sect. 6.3).
(e) Assess risks of incorrect decisions of conformity (Sect. 6.4).
(f) Final assessment of conformity of the entity to specified requirements in terms of impact (Sects. 6.5 and 6.6).

6.3 Response, Restitution and Entropy

6.3.1 Restitution and the Rasch and IRT Models

Decisions are made (at step (d)) on the basis of test results which come from the restitution process which is the final stage of the measurement process for the case of performance metrics and distributions on nominal scales, as illustrated in Fig. 3.3 (which can be compared with Fig. 5.2 of Rossi (2014)) and including the sub-processes—observation and restitution.

In Chap. 2, we described the restitution process of the measurand value z_R, calculated with Eq. (2.9) for the simple example of a measurement system where the instrument sensitivity $K \neq 1$ and there is a quantitative offset (bias), b, in the output:

$$z_{R,j} = S_j = \left(\frac{R-b}{K}\right)_j = \frac{y_j - b_j}{K_j} \tag{2.9}$$

A scheme of the probabilistic model of measurement of Rossi (2014) is given in Fig. 6.2. In addition to the (general) quantities—Z (stimulus) and Y (response)—of the measurement system, the quantity associated with an entity is denoted A (Fleischmann's (1960) *Sachgrösse*; see Chap. 3) which is of course the main focus of this book (and most measurements which are not 'ends in themselves'). The restituted value z_R of the measured object (A) is given, according to probability theory of measurement, by $P(z_R|a) \cdot P(\hat{z}|a);\gamma_R, \gamma$ in the global case.

Alongside a deterministic model of a measurement process (Chap. 2), Rossi with his probabilistic theory of measurement (2014, Sect. 5.3) presents a probabilistic model of the measurement sub-processes including how restitution leads to an estimate Eq. (5.3) of the measurand value z_R in terms of the response, y, of a measurement system to a stimulus, z, in terms of a probabilistic inversion of observation. Invoking the Bayes–Laplace rule, Rossi describes the PMF for the restituted estimate of the measurand as (2014, Eq. (5.45)):

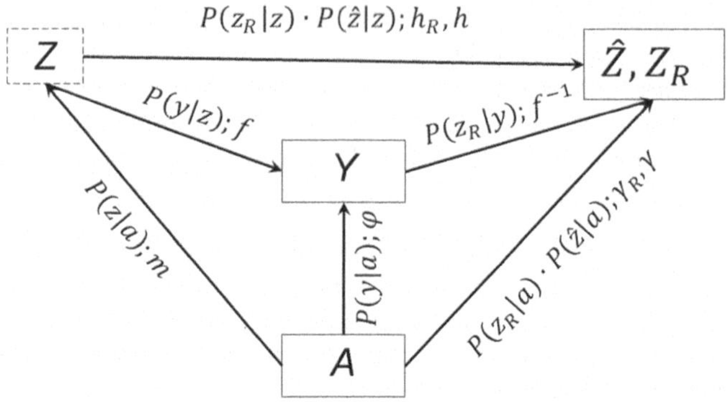

Fig. 6.2 Probabilistic model of the measurement system and processes (adapted from Rossi 2014, Fig. 5.5)

$$P(z_R|y) = \sum_x \left[\frac{P(y|z,x)}{\sum_z P(y|z,x)} \right]_{z=z_R} \cdot P(x) \tag{6.1}$$

In this probabilistic theory of measurement, terms on the RHS of Eq. (6.1) include:

- the PMF $P(y|\, z, x)$ of the response, y, of the measurement system to a stimulus, z, in the presence of nuisance parameters, x,
- a PMF $P(x)$ used to describe the influence on the measurement system of the nuisance parameters, x.

Rossi (2014) illustrates the restitution process in the case of an interfering nuisance parameter, x, with his Fig. 5.9 (comparable with our Fig. 2.6b), which then enters the restitution expression, Eq. (6.8) to yield Rossi's (2014), Eq. (5.40):

$$P(z_R|y,x) = \delta_{\text{Dirac}}\left[z_R - f_x{}^{-1}(z)\right]$$

Once again it has to be pointed out that in many cases the instrument response y is on an ordinal or nominal scale where distances are largely unknown. In such cases it will generally not be possible to formulate the instrument function f or its inverse, as already described in Chap. 2. Restitution in such cases will be difficult where, as plotted in Rossi's Fig. 5.6b—depicting the restitution process $P(z_R|\, y)$; f^{-1} in his probabilistic theory of measurement (Rossi 2014)—the response, y, on the horizontal axis cannot generally be handled as a regular quantitative variable.

In such cases, where the response $y{\sim}P_{\text{success}}$; $P(y|\, a)$; φ is a performance metric (i.e. how accurately classification is made), we have recommended in this book use of logistic regression to transform the ordinal or nominal response to a latent variable description based on expressions such as the 2PL IRT with binary (dichotomous) response:

$$P(y|a);\varphi = P_{\text{success}} = \frac{e^{\rho \cdot (\theta - \delta)}}{1 + e^{\rho \cdot (\theta - \delta)}}$$

where θ and δ are the corresponding performance attributes associated, respectively, with the instrument (probe or person ability) and entity (object or task difficulty). The 2PL expression includes the discrimination, ρ, characteristic of the measurement instrument as a factor additional to these basic Rasch parameters. [We will revisit the question of whether there is some connexion between instrument ability and instrument discrimination in connexion with Eq. (6.9): the less able the instrument, perhaps the greater risk of change (in either direction) of the discrimination.]

Restitution in this categorical response case is then proposed to be based on 2PL IRT, and takes the form:

$$z_R = S = \theta - \delta = \mathit{ln}\left[\frac{P_{\text{success}}}{1 - P_{\text{success}}}\right] - \ln(\rho) \qquad (2.11)$$

6.3.2 *Perceptive Choice and Smallest Detectable Change*

Having obtained the measurement results from the restitution process (Sect. 6.3.1), what kinds of decisions can be based, at step (d), on the test results can now be considered.

Starting with a syntax and semantic approach (Table 3.1) in terms of percentage probabilities of various signs or symbols measured, uncertainty can lead to ambiguity when assessing the significance in general of an apparent difference in pairs of measurement results, for instance, as obtained from two different measurement methods. As shown in Fig. 6.3, two measurement results can be examined as to whether they are significantly different by assessing the distance in entity value separating the two PDF distributions in the corresponding continuous case.

Note that 'uncertainty' here is not a measurement uncertainty in the traditional metrological sense of a standard deviation, but an uncertainty experienced by the decision-maker in the face of limited knowledge when making a choice among the available options.

Two decision scenarios which are found in human-based perception (and analogous system performance metrics), namely: *identification* and *choice*, dealt with in psychophysics (Iverson and Luce 1998; Irwin 2007) serve as models for analogous, measurement-based decisions in other fields. We will deal with *identification* in Sect. 6.5.

Choice involves in the dichotomous case the pairwise discrimination of stimuli, as exemplified in Fig. 6.3. The famous 'Student's' *t*-test involves checking if there is a statistically significant difference between the means of two PDFs $P\left[\overline{z_a}, \frac{(u(z_a))^2}{n_a}\right]$

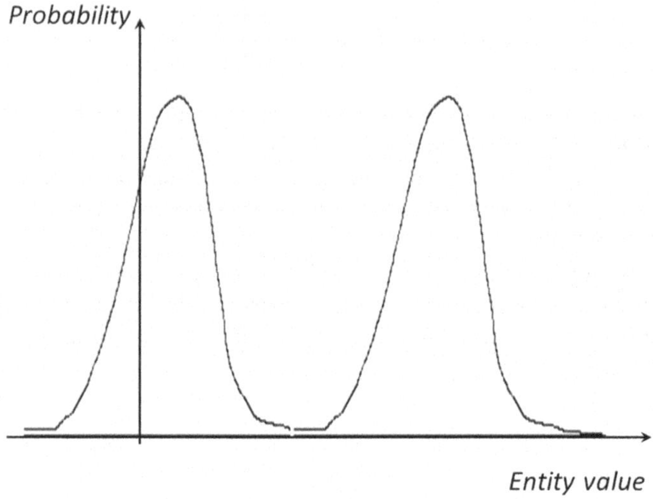

Fig. 6.3 Distance between a pair of test results

and $P\left[\overline{z_b}, \frac{(u(z_b))^2}{n_b}\right]$ compared with their joint standard deviation (e.g. measurement uncertainty, u), according to the formula:

$$|\overline{z_a} - \overline{z_b}| \sim t_{v,95\%} \cdot \sqrt{\frac{(u(z_a))^2}{n_a} + \frac{(u(z_b))^2}{n_b}} \tag{6.2}$$

Any good book of statistics will include a table of Student's t-factors (Table 6.1) for a range of degrees of freedom, v, and statistical significance level (e.g. 95%). For instance, for a series of measurements where the result is a mean of n_a and n_b repeated measurements for the two series a and b, then $v = n_a + n_b - 2$, since the two means have already been determined (i.e. two less degrees of freedom). At 95% significance and $v = 4$, the Student's t-factor is $t_{4.95\%} = 2.87$, which is, so to say, the factor with which a standard uncertainty can be expanded to give a confidence interval with that level of confidence.

This approach can be applied, for instance, in our example of the KCT memory test described in Sect. 5.2, giving an indication of the responsiveness of the test in terms of the various explanatory variables in the CSE Eq. (5.11)

$$X = \{\text{Entropy, Reversals, Average Distance}\}$$

which can be truly detected. The smallest detectable change (SDC) is limited in the first case by 'choice quandary', arising from measurement uncertainties in the psychometric attribute values $u(\delta)$ derived with the RMT Eq. (1.1). A simple derivation of these 'distribution-based' metrics from the general CSE Eq. (5.7) yields the following expression for the SDC associated with the explanatory variable X:

Table 6.1 Student's t-factors

Number of observations	Number of degrees of freedom, ν	Student's 't' factor (double-sided)	k: confidence factor associated with Normal distribution	
n	$n-1$	$p = 31.7\%$ $k = 1$	$p = 5\%$ $k = 2$	$p = 1\%$ $k = 3$
Confidence level, α		$\alpha = 68.3\%$	$\alpha = 95.5\%$	$\alpha = 99.5\%$
2	1	1.84	14.0	–
3	2	1.32	4.53	9.92
4	3	1.20	3.31	5.84
5	4	1.14	2.87	4.60
6	5	1.11	2.65	4.03
7	6	1.09	2.52	3.71
10	9	1.06	2.32	3.25
15	14	1.04	2.20	2.98

$$\mathrm{SDC}(\beta_X \cdot X) = 2 \cdot \sqrt{2} \cdot u(\delta) \Rightarrow \mathrm{SDC}(X) = \frac{2 \cdot \sqrt{2} \cdot \overline{u(\delta)}}{\beta_X} \tag{6.3}$$

On the RHS of this expression:

- The first factor '2' is the coverage factor corresponding to a confidence level of approximately 95% (assuming a Normal distribution).
- The second factor '$\sqrt{2}$' comes from the root-mean-square of the difference between the two distributions being resolved in the SDC, assuming the widths of both distributions are approximately equal.

The responsiveness of each KCT CSE can be judged by evaluating Eq. (6.3) for the two sets of explanatory variables. For the KCT item sequence 2-1-4, entropy corresponds to ~1.8 and the SDC for entropy is 1.5(5). That is, the SDC is smaller than the explanatory variable and, therefore, the explanatory variable entropy is not only the sole explanatory variable of significance, calculated with Eq. (6.3), but also the only explanatory variable that can be truly detected.

The quoted uncertainties (coverage factor $k = 2$) in each SDC have been calculated from the expression for the standard uncertainty:

$$u[\mathrm{SDC}(X)] = \frac{2 \cdot \sqrt{2} \cdot \overline{u(\theta)} \cdot u(\beta_X)}{(\beta_X)^2}$$

where $u(\beta_X)$ is the modelling uncertainty $\boldsymbol{u}(\boldsymbol{\beta}) = \boldsymbol{P} \cdot \boldsymbol{u}(\hat{\boldsymbol{C}})$, provided by the MathCad built-in function *polyfit* used to perform the regression.

The statistical level of significance of a difference can be calculated for a certain difference in means $\overline{z_a} - \overline{z_b}$ as:

$$\alpha_{\text{choice}} = \frac{1}{2} \cdot erfc\left(\frac{|\overline{z_a} - \overline{z_b}|}{\sqrt{\frac{(u(z_a))^2}{n_a} + \frac{(u(z_b))^2}{n_b}}} \right) \tag{6.4}$$

There is, as is well-known, a complete set of statistical significance tests for distributions of individual and average values, as well as tests of variances. These include for variables the *t*-test and Normal tests to determine whether an unknown population mean differs from a standard population mean, and the χ^2-test and *F*-test to determine whether an unknown population standard deviation is greater or less than a standard value (Montgomery 1996).

6.3.3 Significance Testing

As promised earlier (Sect. 2.5.4), if the nominal accuracy of a chosen measurement method has been evaluated in an accuracy experiment, then a number of decisions about test results can be as follows (the SDC example in the previous section is one example):

Choice Between Two Measurement Methods

Two measurement methods may be available when a certain quantity is to be measured. One of the methods is simpler and cheaper than the other, but of less general applicability. Trueness and precision values (Sect. 2.5.2) for each method can then be referred to when motivating the use of the cheaper method for a certain limited range of material where the poorer accuracy is acceptable (ISO 5725-6 1994).

Product Tests Under Conditions of Repeatability

Acceptance testing of measurement results under repeatability conditions can typically be the situation where only a few observations have been made. With only one measurement value, of course, it will be difficult (if not impossible) to draw any conclusions. But with just one additional result, decisions about product can in fact be made, provided a measurement method of known accuracy (repeatability, σ_r) is used. Starting with two results, z_1 and z_2, compare the difference in measurement results with the repeatability limit,[1] $r = 2.8 \cdot \sigma_r$:

[1]The factor '2.8' approximates $2 \cdot \sqrt{2}$ for the root-mean-square of the two results, multiplied by 2 to correspond to a 95% confidence interval (assuming a Normal distribution of the mean), as in the SDC expression Eq. (6.3).

- If $|z_1 - z_2| < r$, then the final result can be readily calculated as $\frac{z_1+z_2}{2}$.
- If $|z_1 - z_2| > r$, then two additional measurements are needed. In that case, the difference in largest and smallest measurement results is compared with the so-called critical range $CR_{0.95}(4) = 3.6$, for 4 measurements (ISO 5725:6 1994):
  - If $(z_{max} - z_{min}) < CR_{0.95}(4)$, then the final result is $\frac{z_1+z_2+z_3+z_4}{4}$.
  - If $(z_{max} - z_{min}) > CR_{0.95}(4)$, then the final result is $\frac{z(2)+z(3)}{2}$

where $z(2)$ and $z(3)$ are, respectively, the second and third smallest measurement results.

Product Tests Under Conditions of Repeatability and Reproducibility

With only one measurement result from each of the two tests made by different setups (measurement systems), one can nevertheless examine whether the absolute difference between the two results exceeds the reproducibility limit: $R = 2.8 \cdot \sigma_R$:

- If acceptable, then the two results are considered to be in conformity and the arithmetic mean value is given as the final result.
- If R is exceeded, then it is necessary to discover whether the difference in measurement results is a result of poor precision in the measurement method and/or a difference in the samples.

Example: Product Specification

A new type of biofuel furnace has been developed and, after a long series of test at the manufacturer, the product is launched on the market. On the furnace datasheet, the manufacturer specifies that the product has been measured to have a combustion efficiency of 45.23%. At a large property consisting of several apartments, the heating system needs renovating. After investigating the market, the property owner chooses the new biofuel furnace. But first he makes a measurement of among others the combustion efficiency. He finds a result of 43.70%. The property owner claims therefore that the manufacturer has overestimated the performance of the furnace since the owner has obtained what seems to be a lower value for the combustion efficiency than that quoted by the manufacturer. Is the difference in result concerning furnace combustion efficiency significantly large?

In both cases (tests by supplier and consumer), one and the same standard measurement method is used which has a known accuracy, since the method has been investigated in an extensive experiment involving several testing laboratories. The accuracy of the method is given in terms of standard deviations under repeatability and reproducibility conditions of, respectively, 0.34% and 0.5%. *Answer*: *Product difference* = 45.23 − 43.70 = 1.53% to be compared with reproducibility limit $R = 2.8 \cdot \sigma_R = 2.8 \cdot 0.5\ \% = 1.4\%$.

That is, there is some evidence for a difference (but more measurements are recommended!).

6.3.4 Significance Testing: Case Study of Pre-packaged Goods

Your name:	Leslie Pendrill
Choose any measurement situation: It can be measurements of the product you have chosen.	Your answers …Coffee powder pre-packaged
• Give an estimate of the precision (scatter) in your measurement method and explain how you have estimated this precision	From section 4.3.1: $\sigma_r =$ $u(z = S) = \mathrm{u}\left(\frac{R-b}{K_{\mathrm{cal}}}\right) = \frac{R}{K_{\mathrm{cal}}} \cdot$ $\sqrt{\frac{u(R-b)^2}{(R-b)^2} + \frac{u(K_{cal})^2}{{K_{\mathrm{cal}}}^2}} = \frac{R}{K_{\mathrm{cal}}} \cdot \frac{u(R-b)}{R-b} = \frac{500\ g}{1} \cdot$ $\frac{1.3\ g}{500\ g-5\ g}$ = 1.3 g, where $u(R-b) =$ $\sqrt{u(R)^2 + u(b)^2} = 1.3\ g$; $u(R) = 0.85$ g from the standard deviation of the mean mass [Table 4.2], at a response level R = 500g of the weighing machine; $u(b)$ = 1 g is the uncertainty on the calibrated bias, b, (typically 5 g) of the weighing machine; and $u(K_{cal}) \sim 0$. It is assumed in this case that $(R - b)$ and K_{cal} are uncorrelated.
Choose two individual measurement results from your measurement data:	$z_1 = 485\ g$ and $z_2 = 488\ g$
• Is the difference between these two results significant compared with the precision of the measurement method? Please give a confidence level (%) in your decision.	Repeatability limit: $r = 2.8 \cdot \sigma_r = 2.8 \cdot 1.3\ g = 3.6\ g$ Observed difference: $\lvert z_1 - z_2 \rvert = 3 < r$ Confidence level 95%. Final result $\frac{z_1+z_2}{2} =$ $486.5\ g$.

6.4 Assessing Entity Error and Measurement 'Value': Cost and Impact

In this section we move on from interpreting a measurement value as merely a technical measure to capture what 'value' the result has in terms of costs and impact. In fact, values can be sought at every level (syntax, semantic, pragmatic, effectiveness) of the quantity calculus hierarchy described in Table 3.1, ultimately including the nature of the quantity and effectiveness, in terms of 'changing conduct' in some active and conscious way.

6.4.1 Uncertainty and Incorrect Estimates of the Consequences of Entity Error

Since each measurement is not perfect, a measurement value is not simply calculated by multiplying a point estimate with the cost at that point, but the effects of measurement uncertainty also have to be vectored in. As illustrated in Fig. 6.4, measurement values will be distributed over the uncertainty interval. Even before assessing conformity to a specification limit (SL_z on entity value, z)—which so to say puts a 'cap' on the costs (Sect. 6.4)—there will in general be an accumulated cost which needs to be integrated across the uncertainty interval wherever it lies. The integral made in each case will depend not only on the uncertainty distribution but also on the cost function, be it linear, u-shaped (Fig. 6.5) or something else. A general expression for risk which pragmatically weighs in costs C together with the probability, p, of an event is

$$\text{Risk} = p \cdot C \tag{6.5}$$

6.4.2 Consequence Costs

Progressing beyond the semantics of percentage probabilities of earlier sections, we adopt a more pragmatic approach to handling decisions about product where measurement 'value' is assessed in terms of costs and impact.

Linearly Priced Goods

A simple example can be taken from our study of pre-packaged goods. Many goods and other commodities (energy, water, environmental emissions, etc.) can

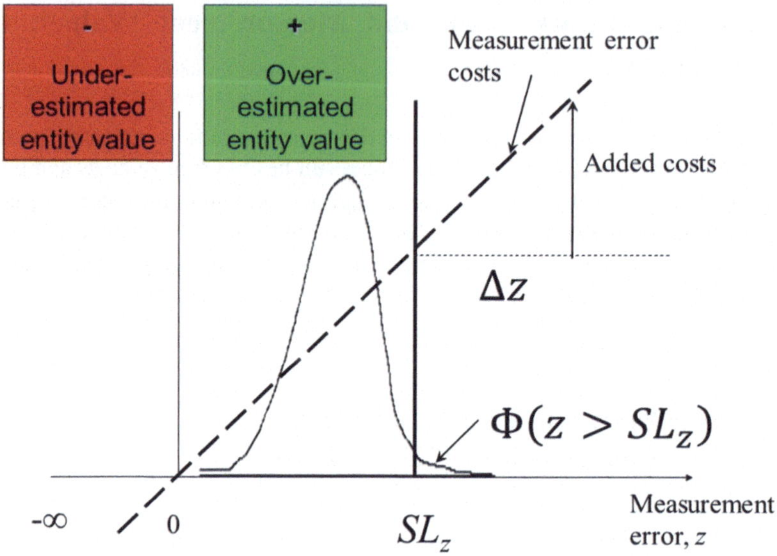

Fig. 6.4 Linear entity costs

be modelled to have simple, linear dependencies of cost or impact value versus the measured quantity value, as illustrated in Fig. 6.4 by the dashed diagonal line, in which impact value $C(z) = K \cdot \Delta z$ increases from zero when the measured value is on target, with the slope, K, of the linear dependency on change in entity value Δz expressed in, for instance, monetary value per measured entity error. Indeed in legal metrology many of these costs are known, and a national economic model can be formulated where the costs of measured error and of providing a legal metrological control can be balanced against income from taxation of goods or environmental emissions as might be required for sustainable development (Pendrill and Källgren 2008; Pendrill 2014a).

Example: Conformity Assessment of Pre-packaged Goods

Referring to your product and measurement demands as well as your measurement data (that you have specified in each section of this document):	Your answers ... *Coffee powder pre-packaged*
(§1.5.2) What are the 'optimal' values of the product's most important characteristics?	*500 g*
(§1.5.1) How large deviations from these optimum values can be tolerated?	*Legal lower specification limit: 485 g*
(§1.5.2) How much will your costs vary with varying deviations in product characteristics?	*Consumer costs of goods: C = 10€/kg*
(§2.2.5) Maximum permissible uncertainty (*MPU*)?	$\mathrm{MPU} = \frac{\mathrm{MPE}}{3} = \frac{15\ g}{3} = 5\ g$
(§6.4.3) How much does the test cost?	*Weighing costs,* $D = \frac{574€}{24h}$ *at an actual, standard uncertainty* $u = 0.6\ g$.
(§6.4.3) What is the real 'value' (e.g. in economic or impact terms) of the measurement values?	*The number of packets manufactured during 24h in the factory studied is typically* 10^5. *Total consumer costs per 24h are therefore 0.5 M€.*

6.4.3 Measurement and Testing Costs

A consideration is how to model the higher cost of increased efforts made to reduce measurement uncertainty, for instance, when seeking an optimised uncertainty (Sect. 6.6). There are of course a number of conceivable models of how test costs could vary with measurement uncertainty.

A common assumption is to assume that the test cost depends inversely on the squared (standard) measurement uncertainty; that is, the test cost is $\frac{D}{u_{\mathrm{measure}}^2}$, where D is the test cost at nominal test (standard) uncertainty u_{measure}. Such a model was suggested (Fearn et al. 2002) based mainly on the argument that N repeated measurements would reduce the standard deviation in the measurement result by $\sqrt{N}$ while costing (at most) N times as much as each individual measurement. In the present work, this model of measurement costs is not only used where the statistical distribution associated with measurement uncertainties is known (such as a type A evaluation), but is also extended to cover more generally even other components of uncertainty (including the expanded uncertainty in the overall final measurement result) where the underlying statistical distribution is often not known (type B evaluation).

In the pre-packed goods example (Sect. 6.4.2), the weighing costs, $D = \frac{57.4€}{24h}$ quoted at an actual, standard uncertainty $u = 0.6$ g were calculated as the sum of the following:

- Each weighing machine is subject to self-checks for a total of 5 h each day, where the checking costs 10€/day.
- Verification of the weighing machine costs 1000€/year.
- Licencing and weighing machine servicing cost an additional 1000€ annually.
- Finally, yearly control visits performed by an external subcontractor cost 700€.

With $N\_day = 10^5$ packets produced each day, and an annual sum of $N\_year = 36.5$ million packets at 500 g each, one can calculate the average weighing cost per day as:

$$D(@u_{\text{measure}}) = 5 \cdot 10 + \frac{1000 + 1000 + 700}{N_{\text{year}}} \cdot N_{\text{day}} = 57.4€/\text{day}$$

6.4.4 *Consumer (Dis-)satisfaction*

The quantity of a product may not be the only concern of the consumer. Where product is valued instead by quality, there is a need to make appropriate models to describe how consumer satisfaction varies with entity value. A common model which aims to account for the fact that the more a product deviates from nominal, the less satisfied will be the consumer is due to Taguchi (1993) as expressed by his loss function: $L_{\text{Taguchi}} = \frac{C}{\left(\frac{U_{\text{SL}} - L_{\text{SL}}}{2}\right)^2} \cdot z^2$, where the consumer risk losses are zero when entity error, *z*, is zero, and equal to *C*, the consumer cost, when entity error, *z*, is at either of the specification limits, U_{SL} or L_{SL}. This is illustrated in Fig. 6.5.

This loss function can replace the linear function shown in Fig. 6.4 when considering the consequences of entity costs when consumer dissatisfaction is a major factor.

Earlier treatments of Taguchi loss functions in conformity assessment include a discussion of the economic setting of guardbands (Williams and Hawkins 1993) and instrument checking intervals (Kacker et al. 1996). A more recent example of this approach can be found in an application to geometrical product control in automobile industry (Pendrill 2010). The practice of guardbanding (Deaver 1994) can also be analysed in terms of costs, impact and optimised uncertainties (Pendrill 2009, Sect. 6.6).

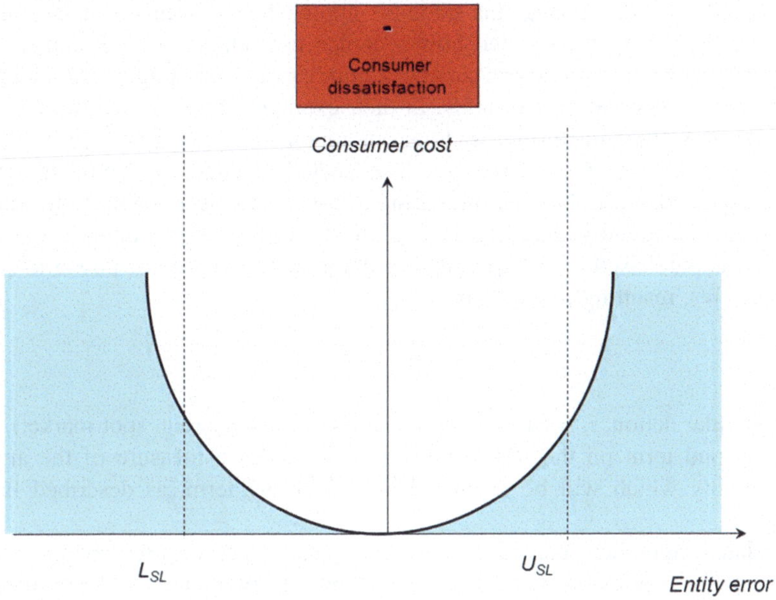

Fig. 6.5 Consumer dissatisfaction costs

6.4.5 'Discrete Choice', 'Prospect' Theory and Costs

A second type of decision-making, apart from choosing between two (or more) alternative signals dealt with above (Sect. 6.3.2), is to weigh measures of utility against risks and uncertainties.

Trading Uncertainty against Utility: Minimal Important Change

When dealing with uncertainty in terms of effectiveness measures of entity value, a 'trade-off' can be made between uncertainty and utility as a means of guiding decisions. An example is according to Yang and Qiu (2005), where a supplier can decide to make a trade-off (coefficient λ) between the expected utility $\mathbb{E}(O_i)$ associated with the outcome O_i of the ith process and the uncertainties in entity economic and quality characteristic, expressed as an entropy term $H(O_i)$. This trade-off is modelled as the risk:

$$R_i = -(1-\lambda)\cdot\frac{\mathbb{E}(O_i)}{\max\mathbb{E}(O_i)} + \lambda\cdot\frac{H(O_i)}{\max H(O_i)} \tag{6.6}$$

An application of making this trade-off approach has been given recently by Gao et al. (2015), who modelled how a choice is made by a large consumer of electrical energy on a smart grid among a number of competing energy suppliers with different risks and cost benefits. In their example, the expected profit, O, for an activity (e.g. manufacturing, trading, communication, providing services, etc.) follows the function of the difference in expected revenue, W, from the activity and the expected price, P, when procuring a product (entity) enabling the activity for a particular quality characteristic, x of the entity. For example, electrical energy (x) is purchased on the spot market at a price (P) to provide power to enable a revenue (W), resulting in a net profit of:

$$O_i = (W_i - P_i) \cdot x_i$$

for a particular action, i, such as a certain supply of energy (e.g. spot market).

The second term on the RHS of Eq. (6.6) includes a measure of the amount of uncertainty which will be evaluated as an entropy term, as described further in Chap. 5.

A related approach considers how the smallest detectable change (SDC) mentioned in connexion with Eq. (6.3) between two signals determined by measurement uncertainty compares with the minimal important change (MIC) as determined in terms, not of measurement (i.e. SDC) but rather, of impact and importance. An example is where the outcome of a measurement is expressed in terms of level of health, such as a Rasch attribute value of patient ability, θ. If MIC is the smallest measured change score that is *perceived* to be important, then when SDC < MIC, it will be possible to distinguish a clinically important change from measurement uncertainty with confidence (van Kampen et al. 2013).

Discrete Choice

The 'discrete choice' approach, such as pioneered by Economy Prize Laureate McFadden (2000), argues that a human being chooses (perhaps subconsciously) to perform a task in a way which maximises her utility, that is, a pay-off between cost and resources mostly in purely economic terms (Ben-Akiva and Bierlaire 1984).

In attempting to model how people choose between various options, the probability, $q_{i,c}$, that person $i = 1,\ldots, N$ chooses category $c = 1,\ldots C$ has been formulated as:

$$q_{i,c} = G\left(x_{i,c}, x_{i,c'} \forall c' \neq c, \theta_i, \beta\right)$$

where the function, G, contains

- vectors, $\boldsymbol{x}$, of attributes for different categories of object,
- vector, $\boldsymbol{\theta}$, of characteristics of person,
- $\boldsymbol{\beta}$—a set of parameters which relates these variables to observed probabilities, often evaluated statistically, for instance, with regression.

Assume that the choice of alternative made by a person seeking to maximise (perhaps subconsciously) utility (or net benefit, well-being, ability, etc.) is given by

$$U_{i,c} = V_{i,c} + \varepsilon_{i,c}$$

Here, the so-called systematic utility is

$$V_{i,c} = \sum_j \alpha_j \cdot X_{j,i,c}$$

where

- α—reciprocal substitution coefficient,
- $X_{j,i,c}$—attribute of path j for person i and choice category, c,
- $\varepsilon_{i,c}$—impact on choice of all unobserved factors (such as other attributes of the system and/or person; measurement and perceptual uncertainties).

This is in accord with expected utility theory, where the expectation is

$$V = \sum_j v_j \cdot p_j$$

as a sum over a number of outcomes of utility v (Economy Prize Laureate Kahneman and Tversky 1979).

An example is the modelling of accessibility:

$$A_{j'} = \sum_{j=1}^{n} O_j \cdot F(C_{j',j})$$

$A_{j'}$—accessibility from zone j'; O_j—number of opportunities in zone j; F—impedance function; $C_{j',j}$—generalised cost from j' to j.

The probability of making a particular choice, c, by maximising utility is then given by

$$q_{i,c} = Pr\left[V_{i,c} + \varepsilon_{i,c} > V_{i,c'} + \varepsilon_{i,c'}\right]; \forall c' \neq c$$

In the log-odds approach:

$$ln\left[\frac{q_{i,1}}{1 - q_{i,1}}\right] = \tau_{i,1} - \beta \cdot z_{i,1}$$

In a multinomial conditional logit approach (Hedeker et al. 2006), with mean = 0 and parameter β, leads in expected utility theory to a choice probability for category c and person i:

$$q_{i,c} = \frac{e^{\beta \cdot z_{i,c}}}{\sum_{c'}^{C} e^{\beta \cdot z_{i,c'}}}$$

which has some obvious similarities to the Lagrange-multiplier expression for the derived response Eq. (5.19).

Prospect Theory

In contrast to traditional expected utility theory, ‘prospect’ theory (Kahneman and Tversky 1979) differs mainly on two points:

- Value is assigned to gains and losses rather than final assets.
- Probabilities are replaced by decision weights.

Decision-making when facing a choice in the presence of risk is made in two stages:

1. *Editing*
   - Decide which outcomes are seen as basically identical.
   - Set a reference point.
   - Consider lesser outcomes as losses and greater ones as gains.
2. *Evaluation*
   - Compute value (utility), based on potential outcomes and their respective probabilities.
   - Choose alternative having higher utility.

The overall or expected utility of outcomes to individual making a (binary) decision is calculated in prospect theory as:

$$V[x, p_x; y, p_y] = v(x) \cdot \pi(p_x) + v(y) \cdot \pi(p_y)$$

in the case of the prospect of making a choice between at most two potential outcomes, x and y, which have probabilities of occurring of p_x and p_y, respectively. The probability of no outcome is $1 - p_x - p_y$.

Decision Weights

In prospect theory, outcomes are weighted with decision weights instead of with the outcome probabilities as done in expected utility theory. The impact on the overall utility associated with the probability, p_x, of outcome x, denoted by a

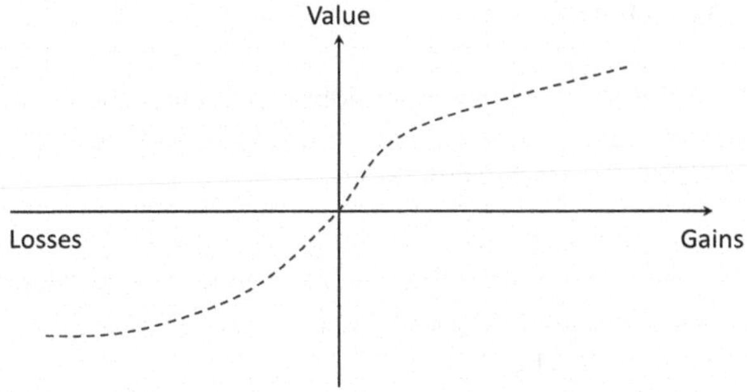

Fig. 6.6 Value as a function of outcome according to prospect theory (Kahneman and Tversky 1979)

decision weight $\pi(p_x)$, as a measure of the desirability, and not merely the perceived likelihood of an event, may typically vary over a range of probabilities.

Prospect theory takes into account that people tend, as seen in empirical studies (Kahneman and Tversky 1979), in some cases to overweight the impact of outcomes of low probability while events of high but finite probability are considered certain.

The subjective value assigned to each outcome is given by the so-called value function, v. This value function (sketched in Fig. 6.6) passes through the reference point, is s-shaped and asymmetrical. Losses hurt more than gains feel good (loss aversion).

6.4.6 Pragmatics and the Rasch Model

In the simplest, dichotomous case with a known prior state, the inclusion of prospect theory decision weights leads to a revised, 'pragmatic' Rasch model, derived as a modified Kullback–Leibler distance as follows:

$$D_{KL,C}(P,Q) = \int -z_c \cdot dP_{\text{success}}$$

$$= -[P_{\text{success}} \cdot C_{\text{win}} \cdot \ln(P_{\text{success}}) + (1 - P_{\text{success}}) \cdot C_{\text{loss}} \cdot \ln(1 - P_{\text{success}})]$$

$$z_c = C_{\text{win}} \cdot \ln\left(\frac{P_{\text{success}}}{1 - P_{\text{success}}}\right) - \Delta C \cdot [\ln(1 - P_{\text{success}}) - 1]$$

where $\Delta C = C_{\text{win}} - C_{\text{loss}}$.

When the consequence costs associated with the different decisions are equal, then the expression returns to the familiar Rasch formula.

Other Models

In addition to the above-mentioned examples, there are various other cost models of product values, for example, concerning company performance or the effectiveness of healthcare (Polo et al. 2005):

- An approach such as discrete choice which focuses mainly on maximising economic utility, while certainly applicable to healthy individuals with the freedom to choose, is felt not to be so relevant when dealing with the chronically ill.
- Product value in finance, for example, is often a collective valuation (e.g. share price) determined by (an often random) influence of many consumers, as in the famous Brownian statistics. As perceived quality varies, associated economic value will sometimes vary abruptly. Again, as in all of the cases considered here, overall costs will be a weighting of value with the probability of a particular entity having a specific value, rather than a purely statistical risk assessment.
- Extreme deviations in entity value from expected can have vanishingly small probabilities of occurrence, but when weighted with an additional, significantly large economic factor, nevertheless have an appreciable impact, as described in terms of a fractal distribution, for instance, see (Mandelbrot and Taleb 2006).

6.5 Comparing Test Result with Product Requirement

A typical test result, with its measurement uncertainty interval, lying in the vicinity of a product specification limit (U_{SL}) is shown in Fig. 6.7 which is an extract of the corresponding Fig. 1.4 as presented from the start in Sect. 1.4.2.

In this section the risks of making incorrect decisions of conformity arising from measurement uncertainty are considered.

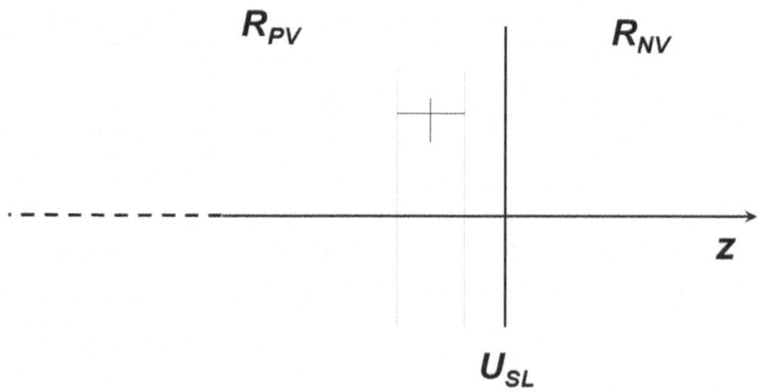

Fig. 6.7 Test result for entity value, z, with its measurement uncertainty interval, lying in the vicinity of a specification limit (U_{SL})

Apart from illustrating product conformity, Fig. 6.7 can also be used when considering measurement conformity. To make decisions about conformity to product specifications (Chap. 1) in a reliable manner will require measurements which themselves have been shown to satisfy conformity to corresponding measurement specifications (Chap. 2). A measurement conformity assessment version of Fig. 6.7 could show, for example, how the actual instrument error (with its uncertainty interval) lies with respect to the maximum permissible (measurement) error (MPE) specification discussed in Chap. 2.

6.5.1 Risks of Incorrect Decisions and Relation to Measurement Uncertainty

Identification (analogous to the psychophysical case (Iverson and Luce 1998)) can involve in the dichotomous case a yes–no detection, as exemplified in Fig. 6.7—is the stimulus within tolerance or outside a region of permissible values R_{PV} specification limit? Is the product correct? Is the perceived shape the correct one? Because of measurement uncertainty, such identifications could be incorrect since, as exemplified in Fig. 6.4, a test result, apparently within limits, might actually be non-conforming since the 'tail' of the probability distribution function extends slightly beyond the limit.

The decision ('consumer') risk, α, i.e. the probability that product is incorrectly approved ('false positive', *FP*) is in this case when sampling by a continuous and quantitative variable as the cumulative distribution function (CDF) beyond the specification limit (U_{SL}, say) on the entity value, z, of the initial set of observations when the mean value $\bar{z}$ is within the limits of the region of permissible values, R_{PV} (JCGM 106 (2012); Pendrill 2014b):

$$\alpha = P(z \geq U_{SL}) = \int_{U_{SL}}^{\infty} \Phi\left[\bar{z} < U_{SL}, u_{\text{measure}}^2\right] \cdot dz$$
$$= \int_{U_{SL}}^{\infty} \frac{1}{\sqrt{2\pi} \cdot u_{\text{measure}}} e^{-\frac{(z-\bar{z})^2}{2 \cdot u_{\text{measure}}^2}} dz \qquad (6.7)$$

assuming a Normal Gaussian probability distribution function $\Phi = N\left(\bar{z}, u_{\text{measure}}^2\right)$. This can be seen mathematically as a specific case of Eq. (6.4) when the width of one of the stimulus distributions is reduced to zero.

Corresponding risks of incorrect decision of conformity when sampling by attribute—that is, where less quantitative results can be sorted in a finite number of discrete categories—can be based on the binomial and Poisson distributions, $g_{\text{attribute}}(p)$ (Joglekar 2003).

Example: Consumer Attribute Risk

A limit, $U_{SL,\,p}$, might be set on the maximum fraction, p, of non-conforming product. Because of measurement (sampling) uncertainty, there is a certain risk that product is non-conforming even though $\hat{p}$, the average fraction non-conforming product lies within specification:

$$Pr(Z > U_{SL,p}|\hat{p}) = \sum_{\eta > U_{SL,p}} g_{attribute}(\eta|\hat{p})\ \hat{p} < U_{SL,p}$$

(The comparison and significance testing of multiple populations can be tackled by conducting analysis of variance (ANOVA) (Joglekar 2003).)

Similar to the other psychophysical case—choice (Sect. 6.3.2), the risks of incorrect decisions on identification—both by variable and by attribute—are simply connected to the logistic regression formula (Eq. (1.1)) by the relation:

$$1 - \alpha = P_{success}$$

6.5.2 *Consumer and Supplier Risks*

Four alternative outcomes—illustrated in Fig. 6.8—can be encountered. In addition to correct decisions of conformity, measurement uncertainty can lead to:

- non-conforming entities being incorrectly passed on inspection—consumer risk, α,
- correctly conforming entities being incorrectly failed on inspection—supplier risk, β,

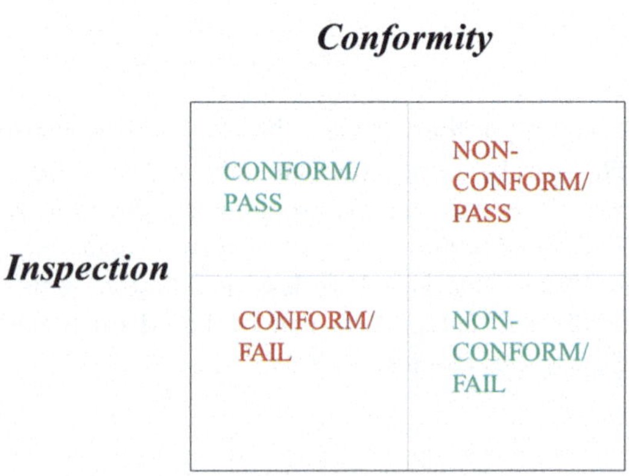

Fig. 6.8 Two correct and two incorrect decisions of compliance

particularly when a test result is close (within the uncertainty interval) to a specification limit.

Making decisions for a single item, the specific risks are:

Consumer specific risk (by variable)

$$\alpha_{\mathrm{specific}}(\bar{z}) = \int_{\eta < L_{\mathrm{SL}}} g_{\mathrm{test}}(\eta|\bar{z}) \cdot d\eta \; \bar{z} \geq L_{\mathrm{SL}}$$

Supplier specific risk (by variable)

$$\beta_{\mathrm{specific}}(\bar{z}) = \int_{\eta \geq L_{\mathrm{SL}}} g_{\mathrm{test}}(\eta|\bar{z}) \cdot d\eta \; \bar{z} < L_{\mathrm{SL}}$$

for a test result mean value, $\bar{z}$, (distribution g_{test}) and a lower specification limit, L_{SL}.

These decision risks can be summarised in a so-called confusion matrix ((Sect. 2.4.4) compared with Fig. 6.8):

$$\begin{pmatrix} 1-\alpha & \alpha \\ \beta & 1-\beta \end{pmatrix}$$

For historical reasons, there are a number of related plots under the name ‘operating characteristic’:

- In statistical acceptance sampling, the ‘operating characteristic’ (or ‘discriminatory power’) curve is a plot of the probability of accepting a lot as a function of the explanatory variable (Pendrill 2008).
- A ‘receiver (or “relative”) operating characteristic’ (ROC) can be a plot of the true positive rate (TPR, or ‘sensitivity’):

$$\mathrm{TPR} = \frac{\mathrm{TP}}{\mathrm{TP}+\mathrm{FN}} = \frac{1-\alpha}{1-\alpha+\beta}$$

against the false positive rate FPR, where β is the ‘supplier’ risk in a dichotomous case, i.e. the probability that product is incorrectly rejected (‘false negative’, FN).

- Irwin (2007) calls plots ‘ROC’ which shows the covariation of success rates for two persons as a function of task difficulty.

Apart from specific risks when assessing conformity to specification of a single item, in general different items will have different quantity values—for instance, due to variations in production or wear and tear on a product—and the commensurate global risks of incorrect decisions also need to be estimated, in line with the terminology and concepts introduced in Sect. 2.2.2. Extensions of the formulae above by convoluting the variables Z_{global} and Y_{global} lead to expressions such as:

Consumer Global Risk (By Variable)

$$\alpha_{\text{global}}(\bar{z}) = \int \int_{\eta < L_{SL}} g_{\text{entity}}(\xi|\bar{z}) \cdot g_{\text{test}}(\eta|\bar{z}) \cdot d\eta \cdot d\xi; \bar{z} \geq L_{SL}$$

Risks and the consequences of incorrect decision-making in conformity assessment should always be evaluated. Beyond the percentage probabilities discussed in this section, ultimately risks can be minimised by proactively setting limits on maximum permissible measurement uncertainties and on maximum permissible consequence costs (Sect. 6.6).

6.5.3 *Mechanistic Model of Binary Decisions: Man as an Operator and Rating the Rater*

In modelling decisions about manufactured product, Akkerhuis (2016), Akkerhuis et al. (2017) combine the traditional statistical decision risk descriptions (Sect. 6.5.2) with a latent variables approach in an attempt at allowing for the rate of correct decision-making by a rater. The basic idea of their work is to modify the calculated decision risk by multiplying the cumulative probability of the measurement result, z, exceeding a specification limit U_{SL}—as expressed traditionally with Eq. (6.7)—with the probability, $P_{\text{success}}[z, \gamma_i]$, of rater i making a 'successful, i.e. correct' rating, as calculated at each value of z. This is illustrated in Fig. 6.9 where the ith rater has a decision threshold, γ_i, interpreted as being the measured product value, z (not necessarily equal to the product specification limit) at which the probability of correct rating is 50% and where items (i.e. product entities) with $z > \gamma_i$ are more

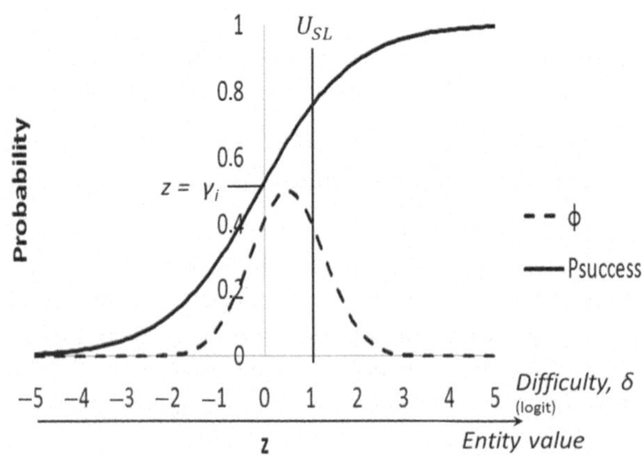

Fig. 6.9 Measurement value PDF distribution, φ, and probability of correct decisions, P_{success}, plotted versus entity value, z, or task difficulty, δ (Rasch rater ability, $\theta = -0, 1$ logit)

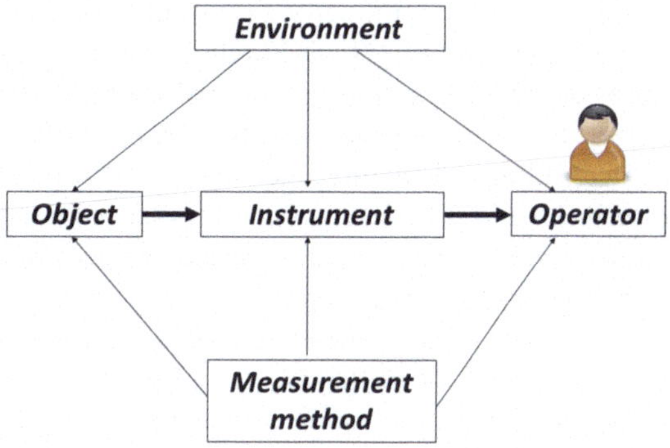

Fig. 6.10 Man as an operator in a measurement system (reproduced from Pendrill 2014a with permission)

likely to be rejected than accepted. Equation (6.7) for the decision risk is modified to become

$$\alpha_i = \int_{U_{\mathrm{SL}}}^{\infty} (1 - P_{\mathrm{success}}[z, \gamma_i]) \cdot \Phi\left[\bar{z} < U_{\mathrm{SL}}, u_{\mathrm{measure}}^2\right] \cdot dz$$
$$= \int_{U_{\mathrm{SL}}}^{\infty} \frac{1}{1 + e^{(z-\gamma_i)}} \cdot \frac{1}{\sqrt{2\pi} \cdot u_{\mathrm{measure}}} e^{-\frac{(z-\bar{z})^2}{2 \cdot u_{\mathrm{measure}}^2}} dz \quad (6.8)$$

(Akkerhuis (2016) also includes a rater discrimination parameter, b (ρ in our notation) which multiplies the exponent in $P_{\mathrm{success}}[z, \gamma_i]$ in a similar manner to the 3PL IRT expression (Eq. (2.10))).

In terms of measurement system analysis where Man enters into different parts of a measurement system (Fig. 1.1), the case tackled by Akkerhuis (2016), Akkerhuis et al. (2017) can be interpreted as Man acting as the Operator of the measurement system, as shown in Fig. 6.10. That is, the operator assesses the accuracy of decisions, based on the response of the instrument, about whether product is approved or not. (Note that this is closely related, but distinct from the case of Man as a measurement instrument (Fig. 1.1b): an operator—who is different from the person acting as an instrument—can make ratings of how well Man as an instrument performs, for example, a teacher rating students.)

In recent work, Andrich (2018) has studied Gaussian and Rasch distributions in instrument response, and van der Bles et al. (2019) consider how epistemic uncertainty is interpreted by the rater, as well as communication about uncertainty to a third-party audience.

We propose that the Rasch approach can usefully be applied to extend the 'rating the raters' approach of Akkerhuis (2016), Akkerhuis et al. (2017): With that extension, separate estimates of the ability, θ_i, of each rater, i, to classify product can be made, alongside estimates of the level of difficulty, δ_j, of each product classification task. To do this, one seeks relations between these Rasch parameters and those of the Amsterdam group, γ_i, z and u_{measure} which enter into Eq. (6.6).

Clearly the item property z is not in itself directly related to either task difficulty, δ, or rater ability, θ. (Akkerhuis (2016) admits that the entity value z is 'unobservable' and U_{SL} ill-defined.) The instrument, which intervenes between object and observer to measure items, will certainly have properties relevant to determining difficulty and ability. Akkerhuis (2016, p. 16) mentions that the discrimination parameter (ρ in our notation) determines the steepness of the characteristic curve of each rater, where a steeper curve corresponds to 'better reliability of the inspection results'.

A solution seems to be formulation of a construct specification equation Eq. (5.7), where the level of difficulty, δ_j, of each product classification task is expressed as a function of a number of explanatory variables, presumably including γ_i, z and u_{measure}. What becomes clear is that, as entity value z varies from below to above the product specification limit U_{SL}, the level of difficulty, δ_j, also varies, while a rater's ability, θ_i, is presumably constant. Well below and well above the specification limit, decisions about entity conformity (identity) are easily made. But close to the specification limit, particularly within a measurement uncertainty, decisions are made with difficulty. A measure of the risk of incorrect decisions as a function of z and u_{measure} is of course expressed by Eq. (6.5) (for the consumer risk case where most of the distribution of z lies below U_{SL}). Instead of convoluting a latent variables expression with the measurement PDF done by Akkerhuis (2016), Akkerhuis et al. (2017) in Eq. (6.6), one can instead equate them:

$$\begin{aligned}\alpha\left[z, \delta_j, \theta_i, \bar{z}, u_{\text{measure}}\right] &= \left(1 - P_{\text{success}}\left[z, \gamma_i\right]\right) = \frac{1}{1 + e^{(z-\gamma_i)}} \\ &= \int_{U_{SL}}^{\infty} \Phi\left[\bar{z} < U_{SL}, u_{\text{measure}}^2\right] \cdot dz \\ &= \int_{U_{SL}}^{\infty} \frac{1}{\sqrt{2\pi} \cdot u_{\text{measure}}} e^{-\frac{(z-\bar{z})^2}{2 \cdot u_{\text{measure}}^2}} dz \end{aligned} \tag{6.9}$$

in accord with our definition of decision-making accuracy—in terms of $P_{success}$, that is, the probability of making a correct categorisation, as in Eq. (2.8):

- $$\textit{Accuracy (decision-making)} = \textit{response categorisation} - \textit{input (true) categorisation} \tag{2.8}$$

Similar observations can be found in the literature: In interpreting the slope parameter, ρ, of the 2PL IRT model, earlier workers (Uebersax and Grove 1993) quote the relation $\sigma = \frac{1.7}{\rho}$ obtained by 'approximating the logistic ogive with the Gaussian cumulative distribution function with an appropriately chosen mean and standard deviation (σ)' (Petersen et al. 2012). The term '1.7' can be understood as $\sqrt{3}$ as it enters in the expression for the entropy $H(P, Q) \sim \ln(\rho) = ln\left(\sqrt{3} \cdot 2 \cdot u\right)$ of a uniform distribution (Fig. 4.4) associated with the finite resolution, ρ, of an instrument, where u is the standard measurement uncertainty, as described in Sect. 5.4.1.

As pointed out earlier at several places, the Rasch psychometric approach is not restricted to measurement systems with human intervention, but should apply more generally to other 'probe: task' systems (illustrated in Fig. 4.11 and in Sect. 5.4.2). Examples include (Pendrill 2014a, b; Turetsky and Bashkansky 2015):

- the performance of a system (characterised by the ability of providing healthcare, such as waiting times for surgery, separately from the levels of challenge associated with each task),
- the determination of material testing (e.g. the ability of an indenter, separately from the hardness of each material test block).

6.5.4 Multivariate Decision-Making

All of the discussion this far has addressed a univariate decision-making case, as illustrated in Fig. 6.7, where a single entity value dimension, *z*, has a specification limit. There remain in the international literature a number of different approaches to defining multivariate specification regions. One example is engineering specifications for the PC_is and their target values in multivariate process control proposed in the approach of Wang and Chen (1998) as: $LSL_{PC_i} = p_i^T \cdot LSL$; $USL_{PC_i} = p_i^T \cdot USL$, with a corresponding (rectangular) multivariate PC specification region (Fig. 6.11).

All of these different approaches assume a linear model and quantitative variables. Accounts of non-linear multivariate studies can be found in the literature, while the use of GLM in compensating for ordinal data (e.g. for counted fractions—(Chap. 3)) has only recently entered the field (Pendrill et al. 2015).

6.6 Optimised Uncertainties, Impact and Measurement Costs, Pragmatic Extensions of Significance Testing

In general, the impact of a wrong decision in conformity assessment is expressed as a Risk, defined as the probability *p* of the incorrect decision occurring multiplied by the cost *C* of the consequences of the incorrect decision (AFNOR 2004):

$$\text{Risk} = p \cdot C \tag{6.5}$$

As promised in Chap. 2, questions of appropriate rules for decision-making in conformity assessment with due account of measurement uncertainty raise questions about fit-for-purpose measurement (Sect. 2.1) which ultimately can be resolved by economic considerations.

Indeed, revisiting our hierarchy of concepts in information theory and quantity calculus (Table 3.1) including costs in the risk according to Eq. (6.5) is at the pragmatic level. In recent work, van der Bles et al. (2019) have considered the communication of epistemic uncertainty to both the rater (Fig. 6.10) as well as

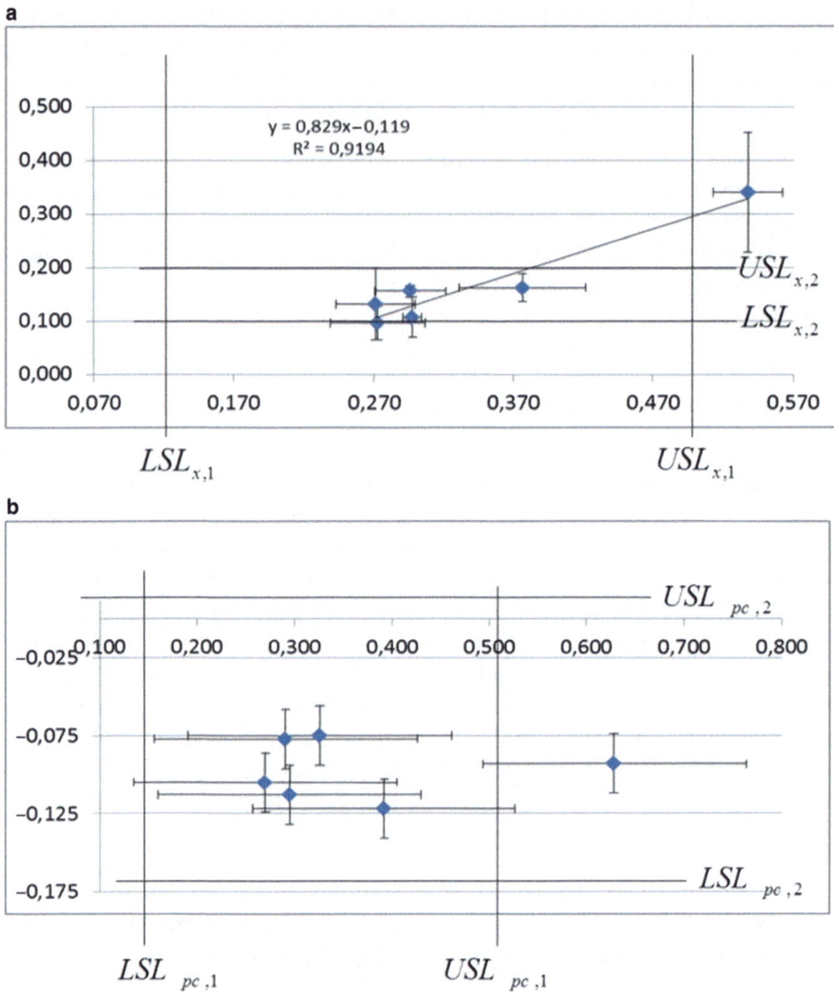

Fig. 6.11 Bivariate specification regions and decision-making: (**a**) by explanatory variable Z for the specification region $S_z = \{z \in Z | LSL_z \leq z_i \leq USL_z\}$ and (**b**) by principal component, *pc*, for the specification region $S_{PC} = \{z'_i \in Z' | LSL_{PC} \leq z'_i \leq USL_{PC}\}$ (Pendrill et al. 2015)

third-party audiences where the 'goal of communication is to affect an audience in some way: to inform, motivate, instruct or influence people'. That we would consider to lie at the highest, effectiveness level (Table 3.1), that is, 'changing conduct': relationship between signs of communication and actively improving the 'entities' the signs stand for (Weinberger 2003).

The traditional 'confusion matrix' of statistical hypothesis testing (Fig. 6.8) is augmented (Fig. 6.12) with an economic (loss function or more generally impact, as well as income) approach. Instead of arbitrary percentage risks, the decision-maker can assess real costs—be they profits or losses.

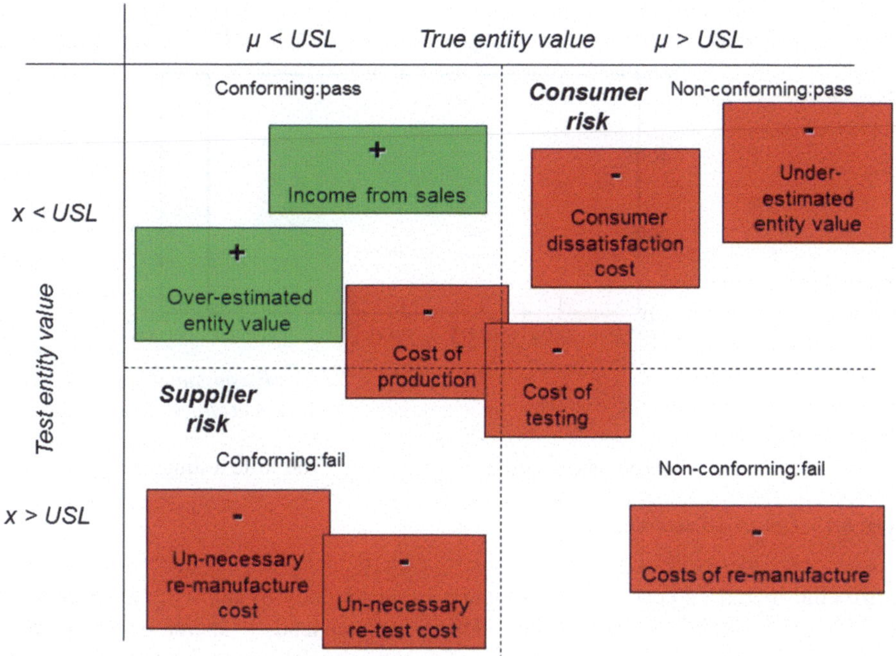

Fig. 6.12 Confusion matrix with costs and incomes added (Reproduced with permission from Figure 1 of Pendrill 2014b)

Consumer risk cost (lower specification limit, LSL):

$$C_{\text{specific}}(\bar{z}) = \int_{\eta < L_{\text{SL}}} C(\eta) \cdot g_{\text{test}}(\eta|\bar{z}) \cdot d\eta; \bar{z} \geq L_{\text{SL}}$$

Supplier risk cost (lower specification limit, L_{SL}):

$$C^*{}_{\text{specific}}(\bar{z}) = \int_{\eta \leq L_{SL}} C^*(\eta) \cdot g_{\text{test}}(\eta|\bar{z}) \cdot d\eta; \bar{z} < L_{\text{SL}}$$

The consequence costs, C, of incorrect decisions for test result mean (restituted) value, $\bar{z}$, (distribution g_{test}) can be balanced against measurement and test costs in the optimised uncertainty approach (Thompson and Fearn 1996; Pendrill 2014b), where the costs of measurement (exemplified in Sect. 6.4) are modelled as $\frac{D}{u_{\text{measure}}^2}$, where D is the test cost at nominal test (standard) uncertainty u_{measure}.

One seeks a minimum at an optimised uncertainty in the overall economic value, E, which is the sum of measurement and consequence costs, for instance, involving consumer risks:

$$E(u_{\text{measure}}|\bar{z}) = \frac{D}{u_{\text{measure}}^2} + C_{\text{specific}}(\bar{z}) \tag{6.10}$$

Figure 6.13 shows an idealised plot of the optimised uncertainty methodology according to Eq. (6.10). Reducing measurement uncertainty leads to less decision

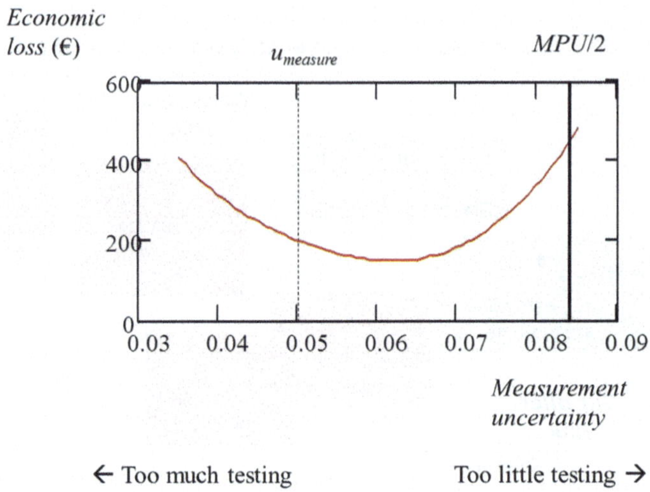

Fig. 6.13 Optimised uncertainty

errors but increased costs of testing, while saving money by allowing uncertainties to increase by doing too little testing sooner or later leads to increased costs from the consequences of incorrect decisions of conformity. The optimised uncertainty is thus a kind of 'golden mean' and, as shown in Fig. 6.13, is often a more motivated, 'fit-for-purpose' choice compared with traditional 'rule-of-thumb' estimates such as $\frac{\mathrm{MPU}}{2}$ in terms of more arbitrary maximum permissible uncertainty, MPU, limits (Sect. 2.2.4).

Equation (6.10) can be evaluated in two complementary ways as illustrated in Fig. 6.14. In addition to the optimised uncertainty plot of Fig. 6.13 (a range of test uncertainties, u_{measure}, for a given mean entity value $\bar{z} < U_{\mathrm{SL}}$), one can make a complementary plot:

• a range of quantity values of $U_{\mathrm{SL}} - h \cdot u_{\mathrm{measure}} \leq \bar{z} \leq U_{\mathrm{SL}} + h \cdot u_{\mathrm{measure}}$ for a given test uncertainty, u_{measure}, and 'guardband' factor h—yielding an 'operating cost characteristic' which extends the traditional concept of a probability-based operating characteristic to include the effects of impact (Pendrill 2008).

A decision theory approach balances pragmatically the costs of analysis against the costs associated with the consequences of incorrect decision-making. A general expression for the total cost, E, associated with a test procedure is the sum of costs of testing and the consequences of incorrect decisions:

$$E = D_{\mathrm{measure}} + \iint C_{\mathrm{consequence}}(z, \mu) \cdot P_{\mathrm{measure}}(z|\mu) \cdot P(\mu) \cdot dz \cdot d\mu \tag{6.11}$$

where the cost, D_{measure}, of taking samples and making tests has to be balanced against the possible losses $C_{\mathrm{consequence}}$ when the test results are used in decision-making. Expression (6.11) allows for dispersion P (PDF) in the measurement results z as well as in the entity value μ, that is, in line with the terminology and concepts

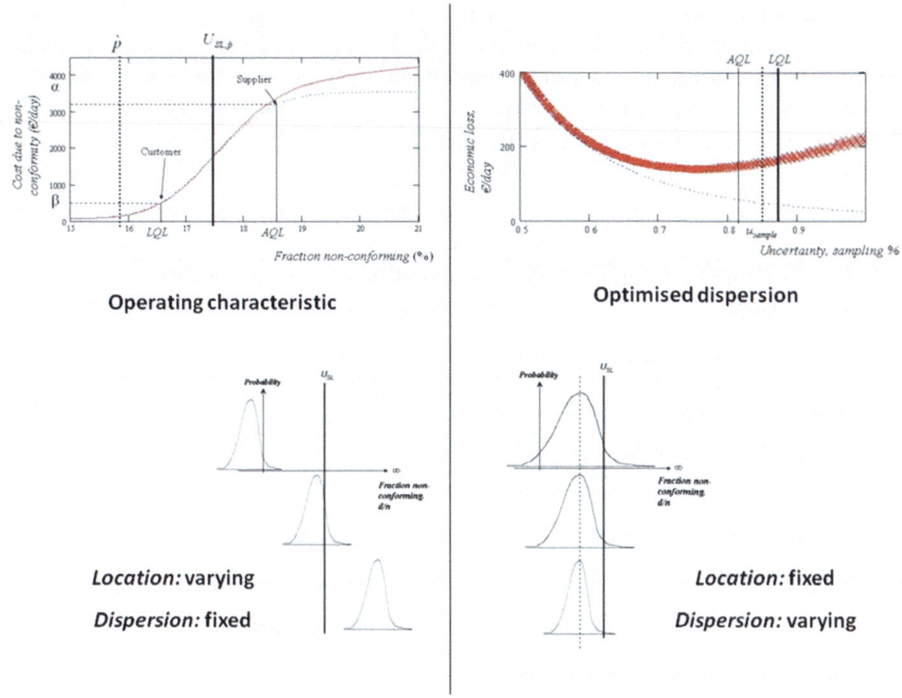

Fig. 6.14 Operating cost characteristic and optimised dispersion (Reproduced with permission from Figure 2 of Pendrill 2014b)

introduced in Sect. 2.2.2 for specific and global risks. Making calculations of the total costs with Eq. (6.11) allowing for dispersions in both directions—measurement and product—enables a 3D plot to be made (with E on the vertical axis, as illustrated in Fig. 9 of Pendrill 2008) in which an optimised uncertainty can be sought at a 'valley' minimum somewhere across the three-dimensional surface.

$$E_{\mathrm{np}} = \frac{D_{\mathrm{np}}}{u_{\mathrm{measure}}^{2}} + C_{\mathrm{np}} \cdot \left[\int_{-\infty}^{\mathrm{USL}_z} \frac{1}{\sqrt{2 \cdot \pi} \cdot s_{\mathrm{instrument}}} \cdot e^{-\frac{(z - z_{\mathrm{instrument}})^2}{2 \cdot s_{\mathrm{instrument}}^{2}}} \cdot \int_{\mathrm{USL}_x}^{\infty} \frac{1}{\sqrt{2 \cdot \pi} \cdot u_{\mathrm{measure}}} \cdot e^{-\frac{(x - x_{\mathrm{measure}})^2}{2 \cdot u_{\mathrm{measure}}^{2}}} \cdot dx \cdot dz \right] \tag{6.12}$$

Various cost models can be employed: linear models for metering in legal metrology, for instance, parabolic functions capturing varying consumer satisfaction of market expectations, etc. (Taguchi 1993, Fig. 6.3). Studies of pre-packaged goods (Pendrill 2008, next section) in legal metrology are a kind of prototype for a general treatment of (univariate, interval scale) conformity assessment of any kind of product. A more recent example of this approach can be found in an application to geometrical product control in automobile industry including consumer dissatisfaction (Pendrill 2010). The practice of guardbanding (Deaver 1994), which attempts to minimise the risks of incorrect decisions associated with measurement uncertainty by surrounding a product specification limit by a 'protective' interval which is some multiple of the uncertainty, can also be analysed in terms of costs, impact and optimised uncertainties (Pendrill 2009).

6.6.1 Example: Conformity Assessment of Pre-packaged Goods

Referring to your product and measurement demands as well as your measurement data (that you have specified in each section of this document):	Your answers …Coffee powder pre-packaged
(§4.3.1) From your measurement results:	
Is your actual measurement uncertainty within measurement specification (i.e., *MPU*)?	Actual, standard uncertainty $u = 0.6g$ is well within $MPU = 5g$
Is the test result (including uncertainty interval) within product specification (i. e., *MPE*) about the 'optimum' product value? Is the product approved or not?	
Give the measurement uncertainty and test result location with respect to product specification limits; risks for erroneous decisions when assessing compliance ('conformity assessment') Express these preferably in terms of consumer and supplier risks, either in % or preferably in tangible terms (e.g. economy)	
Does the actual measurement uncertainty lie close to the 'optimum' uncertainty, i.e. after having balanced (§2.1) measurement- and (§1.2) consequence costs?	See Fig. 6.13
When you communicate your results to the task assigner, what will be your last words?	
• Others:	

The results of an optimised uncertainty analysis for the case of pre-packaged coffee are shown in Fig. 6.15.

Typical results for the case of pre-packaged coffee are listed in Table 6.2.

76/211/EEC,
75/106/EEC

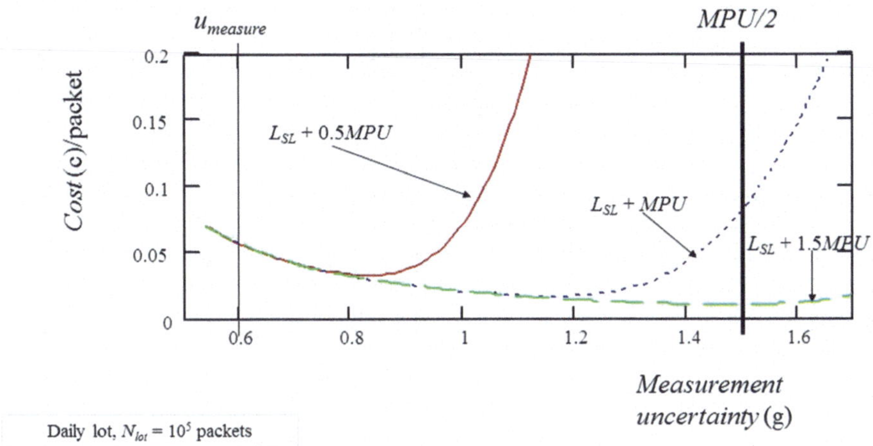

Fig. 6.15 Optimised uncertainty analysis for the case of pre-packaged coffee

6.6.2 *Optimised Uncertainties on an Ordinal Scale*

In addition to introducing a cost function when determining the *effectiveness of sorting*, that is, how dispersed the classification is, on an ordinal scale for an entity of given reference level, we can also proactively specify (step 0) a maximum permissible uncertainty as well as assess the consequences of measurement uncertainty on incorrect decisions of product conformity, as follows:

For any given task, *an optimised measurement uncertainty* can be identified by modelling the overall (economic) impact of appraisal as the sum of the consequence costs (expected loss, $\mathrm{EL} = \sum_j C_j \cdot q_j$) balanced against the costs of measurement. The lower the measurement uncertainty, i.e. the better the ability of the person to classify, the more it costs to make the measurement—for instance, owing to the extra cost of educating, training (and paying!) the rater. At the same time, a lower measurement uncertainty will have lower consequence costs for incorrect decisions. The optimum measurement uncertainty is that corresponding to a minimum in the overall cost of appraisal as a function of uncertainty.

Supplier and consumer risks with limited sampling (sample size N_{sample} taken from a lot of size N_{lot}) can be estimated in terms of the 'tail' in the sampling distribution, $g_{\mathrm{attribute}}$, beyond the specification limit $\mathrm{USL}_{\hat{p}}$ on fraction non-conforming entities $\hat{p} = \frac{d}{N_{\mathrm{sample}}}$:

Table 6.2 Measurement and product values for the case of pre-packaged coffee

Product values			Measurement	Costs		Sampling	
Mean mass, $\bar{z}$/packet	Standard deviation, σ_p	Specification mass, $L_{SL,\bar{z}}$	Measurement uncertainty, mass, $u_{measure}$	Measurement cost, D	Consequence cost, $C_{specific}$	Daily lot, N_{lot}	Sample size, N_{sample}
493 g	8 g	485 g	0.6 g	57.4 €/day	10 €/kg	10^5	2000

$$Pr\left(Z > \mathrm{USL}_{\hat{p}} | \hat{p}\right) = \sum_{\eta > \mathrm{USL}_{\hat{p}}} g_{\text{attribute}}(\eta | \hat{p})$$

The cost associated with the probability $Pr\left(Z > \mathrm{USL}_{\hat{p}} | \hat{p}\right)$of the fraction, $\hat{p} = \frac{d}{N_{\text{sample}}}$, non-conforming product, where Z lies outside a specification limit, but is passed on inspection since $\hat{p}$ lies inside the region of permissible values—constituting a consumer's attribute risk, is evaluated as:

$$C_{\text{test}}(\hat{p}) = \sum_{p > \mathrm{USL}_{\hat{p}}} C_{\text{survey}}\left(N_{\text{lot}}, C_{P,S}, \hat{z}(p), s_p\right) \cdot \frac{p}{\hat{p}} \cdot g_{\text{test}}(p | \hat{p})$$

$$\tilde{\ } C_{\text{survey}}\left(N_{\text{lot}}, C_{P,S}, \hat{z}, s_p\right) \cdot \sum_{p > \mathrm{USL}_{\hat{p}}} g_{\text{test}}(p | \hat{p}) \qquad (6.13)$$

with $\hat{p} \leq \mathrm{USL}_{\hat{p}}$ in the case of an upper specification limit, $\mathrm{USL}_{\hat{p}}$, where

- $C_{\text{survey}}\left(N_{\text{lot}}, C_{P,S}, \hat{z}, s_p\right)$—the (approximately constant) cost of product exceeding a (lower) specification limit on measured (restituted) quantity, z, with dispersion s_p, and unit cost $C_{P,S}$,
- $g_{\text{test}}(p | \hat{p})$—the distribution of fraction non-conforming product—is given by Eq. (6.13) in the case of sampling from an infinitely large lot (Pendrill 2008).

Normally, a person's ability is not based on a single, specific test but rather on an aggregate over a range of tasks of different levels of challenge. The *operating 'cost' characteristic* approach, together with the optimised uncertainty methodology, would then provide a three-dimensional mapping of how overall appraisal costs vary at different levels of challenge and measurement uncertainty.

Exercises: Measurement and Product Decisions

Conformity Assessment

Referring to your product and measurement demands as well as your measurement data (that you have specified in each section of this document)	Your answers...
(§1.2) What are the 'optimal' values of the product's most important characteristics?	
(§1.2) How large deviations from these optimum values can be tolerated?	

(continued)

Referring to your product and measurement demands as well as your measurement data (that you have specified in each section of this document)	Your answers. . . . .
(§1.2) How much will your costs vary with varying deviations in product characteristics?	
(§2.2) Maximum permissible uncertainty (*MPU*)?	
(§2.1) How much does the test cost?	
What is the real 'value' (e.g. in economic or impact terms) of the measurement values?	
(§4.2) From your measurement results	
• Is your actual measurement uncertainty within measurement specification (i.e. *MPU*)?	
• Is the test result (including uncertainty interval) within product specification (i.e. *MPE*) about the 'optimum' product value? Is the product approved or not?	
• What is the real 'value' (e.g. in economic or impact terms) of the measurement values?	
• Give the measurement uncertainty and test result location with respect to product specification limits; risks for erroneous decisions when assessing compliance ('conformity assessment'). Express these preferably in terms of consumer and supplier risks, either in % or preferably in tangible terms (e.g. economy)	
• Does the actual measurement uncertainty lie close to the 'optimum' uncertainty, i.e. after having balanced (§2.1) measurement and (§1.2) consequence costs?	
When you communicate your results to the task assigner, what will be your final words?	
Others:	

Significance Testing

Choose any measurement situation: It can be measurements of the product you have chosen	Your answers..
• Give an estimate of the precision (scatter) in your measurement method and explain how you have estimated this precision	
Choose two individual measurement results from your measurement data	
• Is the difference between these two results significant compared with the precision of the measurement method? Please give a confidence level (%) in your decision	
Others:	

References

AFNOR, Use uncertainty in measurement: presentation of some examples and common practices, in *French Standardisation* FD x07–022 (2004)

T. Akkerhuis, Measurement system analysis for binary tests, PhD thesis, FEB: Amsterdam Business School Research Institute (ABS-RI) (2016). ISBN 9789462333673. http://hdl.handle.net/11245/1.540065

T. Akkerhuis, J. de Mast, T. Erdmann, The statistical evaluation of binary test without gold standard: Robustness of latent variable approaches. Measurement **95**, 473–479 (2017). https://doi.org/10.1016/j.measurement.2016.10.043

D. Andrich, On an identity between the Gaussian and Rasch measurement error distributions: Making the role of the instrument explicit. J. Phys. Conf. Series **1065**, 072001 (2018). https://doi.org/10.1088/1742-6596/1065/7/072001

M. Ben-Akiva, M. Bierlaire, Discrete choice methods and their application to short term travel decisions, in *International Series in Operations . . ., 1999* (1984). books.google.com

D. Deaver, *Guardbanding with confidence Proc. NCSL Workshop & Symposium*, Chicago, July–August 1994 (1994), pp. 383–394

T. Fearn, S. A. Fisher, M. Thompson and S. Ellison, A decision theory approach to fitness for purpose in analytical measurement Analyst **127** 818–24 (2002)

R. Fleischmann, Einheiteninvariante Gröβengleichungen, Dimension. Der Mathematische und Naturwissenschaftliche Unterricht **12**, 386–399 (1960)

B. Gao, C. Wu, Y. Wu and Y. Tang, "Expected Utility and Entropy-Based Decision-Making Model for Large Consumers in the Smart Grid", Entropy 17, 6560-6575; doi:10.3390/e17106560 (2015)

D. Hedeker, M. Berbaum, R. Mermelstein, Location-scale models for multilevel ordinal data: between- and within subjects variance modeling. J Probab. Stat. Sci. **4**(1), 1–20 (2006)

R.J. Irwin, A psychophysical interpretation of Rasch's psychometric principle of specific objectivity, in *Proceedings of Fechner Day*, **23** (2007).

G. Iverson and R. Luce, The representational measurement approach to psychophysical and judgmental problems, in Measurement, Judgment, and Decision Making. Academic Press Cambridge (1998)

ISO 5725, *Accuracy (trueness and precision) of measurement methods and results, Part 6: Use in practice of accuracy values* (1994)

JCGM 106:2012, Evaluation of measurement data – The role of measurement uncertainty in Conformity Assessment, in *Joint Committee on Guides in Metrology (JCGM)* (2012)

A.M. Joglekar, *Statistical Methods for Six Sigma in R&D and Manufacturing* (Wiley, Hoboken, 2003). ISBN: 0-471-20342-4

R. Kacker, N.F. Zhang, C. Hagwood, Real-time control of a measurement process. Metrologia **33**, 433–445 (1996)

D. Kahneman, A. Tversky, Prospect theory: An analysis of decision under risk. Econometrica **47**(2), 263–291 (1979). http://www.princeton.edu/~kahneman/docs/Publications/prospect_theory.pdf

B. Mandelbrot, N Taleb, Wild uncertainty, in *Financial Times,* 2006-03-24, Part 2 of 'Mastering Uncertainty' series (2006)

D.L. McFadden, Economic choices, in *Prize Lecture*, Stockholm (SE), 8 December 2000 (2000). http://www.nobelprize.org/nobel_prizes/economics/laureates/2000/mcfadden-lecture.pdf

D.C. Montgomery, *Introduction to Statistical Quality Control* (Wiley, Hoboken, 1996). ISBN: 0-471-30353-4

L.R. Pendrill, Operating 'cost' characteristics in sampling by variable and attribute. Accred. Qual. Assur. **13**, 619–631 (2008)

L.R. Pendrill, "An optimised uncertainty approach to guard-banding in global conformity assessment", *Advanced Mathematical and Computational Tools in Metrology VIII,* in *Data Modeling for Metrology and Testing in Measurement Science Series: Modeling and Simulation in Science, Engineering and Technology, Birkhauser,* Boston 2009. ISBN: 978-0-8176-4592-2. http://www.worldscibooks.com/mathematics/7212.html

L.R. Pendrill, Optimised uncertainty and cost operating characteristics: new tools for conformity assessment. Application to geometrical product control in automobile industry. Int. J. Metrol. Qual. Eng **1**, 105–110 (2010). https://doi.org/10.1051/ijmqe/2010020

L.R. Pendrill, Man as a measurement instrument. NCSLI Measure J. Meas. Sci. **9**, 24–35 (2014a)

L.R. Pendrill, Using measurement uncertainty in decision-making & conformity assessment. Metrologia **51**, S206 (2014b)

L.R. Pendrill, H. Källgren, Optimised measurement uncertainty and decision-making in the metering of energy, fuel and exhaust gases. Izmerite'lnaya Technika (Meas. Tech.) **51**(4), 370–377 (2008). https://doi.org/10.1007/s11018-008-9047-8

L.R. Pendrill, H. Karlsson, N. Fischer, S. Demeyer, A. Allard, A guide to decision-making and conformity assessment, in *Deliverable 3.3.1, EMRP project (2012–5) NEW04 Novel Mathematical and Statistical Approaches to Uncertainty Evaluation* (2015). http://www.ptb.de/emrp/new04-publications.html

J.H. Petersen, K. Larsen, S. Kreiner, Assessing and quantifying inter-rater variation for dichotomous ratings using a Rasch model. Stat. Methods Med. Res. **21**, 635–652 (2012). https://doi.org/10.1177/0962280210394168

F.-J.V. Polo, M. Negrin, X. Badia, M. Roset, Bayesian regression models for cost-effectiveness analysis. Eur. J. Health Econ. **1**, 45–52 (2005)

G.B. Rossi, Measurement and probability – A probabilistic theory of measurement with applications, in *Springer Series in Measurement Science and Technology*, (Springer, Berlin, 2014). https://doi.org/10.1007/978-94-017-8825-0

G. Taguchi, *Taguchi on Robust Technology* (ASME Press, New York, 1993)

M. Thompson, T. Fearn, What exactly is fitness for purpose in analytical measurement? Analyst **121**, 275–278 (1996)

V. Turetsky, E. Bashkansky, Testing and evaluating one-dimensional latent ability. Measurement **78**, 348–357 (2015)

J.S. Uebersax, W.M. Grove, A latent trait finite mixture model for the analysis of rating agreement. Biometrics **49**, 823–835 (1993)

A.M. van der Bles, S. van der Linden, A.L.J. Freeman, J. Mitchell, A.B. Galvao, L. Zaval, D.J. Speigelhalter, Communicating uncertainty about facts, numbers and science. R. Soc. Open Sci. **6**, 181870 (2019). https://doi.org/10.1098/rsos.181870

D.A. van Kampen, W.J. Willems, L.W.A.H. van Beers, R.M. Castelein, V.A.B. Scholtes, C.B. Terwee, Determination and comparison of the smallest detectable change (SDC) and the minimal important change (MIC) of four-shoulder patient-reported outcome measures (PROMs). J. Orthop. Surg. Res. **8**, 40 (2013). http://www.josr-online.com/content/8/1/40

F.K. Wang, J.C. Chen, Capability index using principal component analysis. Qual. Eng. **11**, 21–27 (1998)

E.D. Weinberger, A theory of pragmatic information and its application to the quasi-species model of biological evolution. Biosystems **66**, 105–119 (2003). http://arxiv.org/abs/nlin.AO/0105030

R.H. Williams, C.F. Hawkins, The Economics of Guardband Placement, in *Proceedings, 24th IEEE International Test Conference,* Baltimore, MD, 17–21Oct. 1993 (1993)

J. Yang, W. Qiu, A measure of risk and a decision-making model based on expected utility and entropy. Eur. J. Oper. Res. **164**, 792–799 (2005)

Index

L. Pendrill, *Quality Assured Measurement*, Springer Series in Measurement Science and Technology, https://doi.org/10.1007/978-3-030-28695-8

The manufacturer's authorised representative in the EU is Springer Nature Customer Service Centre GmbH, Europaplatz 3, 69115 Heidelberg, Germany. If you have any concerns regarding our products, please contact ProductSafety@springernature.com

Printed and bound by CPI Group (UK) Ltd, Croydon, CR0 4YY
05/10/2026
02228130-0006